Matlab Technology Resource Manual

Herman Gollwitzer
Drexel University

---

to accompany

Contemporary Linear Algebra

D0223250

Howard Anton
Drexel University

Robert C. Busby
Drexel University

WILEY

To order books or for customer service call 1-800-CALL-WILEY (225-5945).

ISBN 0-471-26940-9

Printed in the United States of America

10 9 8 7 6 5 4 3 2 1

Printed and bound by Hamilton Printing Company

Introduction

The manual consists of explanations of the technology exercises using MATLAB. Problem solving involves formulation, computation, and interpretation. Computations in MATLAB are organized in text files called M-files, and they are an essential part of MATLAB practice. Most exercises have an associated M-file that generates the data, carries out the computations, and presents the results. Over two hundred in number, they are called Learning Codes. Reading through the Learning Codes is a good way to acquire an understanding of MATLAB syntax and commands and to see connections between linear algebra and computation. Elementary MATLAB commands are introduced in context, and the programming style is designed for readability rather than maximum efficiency. Extensive comments accompany the computations. Being text files, Learning Codes provide an extensible environment for linear algebra computations. In several cases, the M-files are written so that similar problems can be explored with only minor changes. Many Learning Codes provide a working platform for exploration. Visualization is used for emphasis, and the MATLAB Symbolic Toolbox is employed where appropriate. Learning Codes are to be read, run, understood, taken apart, criticized, and hopefully improved. Doing so can be part of class projects or other activities.

Getting familiar with the user interface depends on which version you are using. Students should look to their instructors for guidance in this matter. Later versions have animated help sessions that indicate how the Command Window is managed. Go to the help menu. Of greatest importance is an understanding of the current directory concept. At any given time, there is a directory called the current directory that should contain all M-files of immediate interest. When MATLAB is launched, a current directory is assigned in a manner that is system dependent. It is probably not the directory (folder) where your files reside. The current directory is displayed and set in the upper right corner of the Command Window in recent versions. A popup menu gives recent directories, and a button next to it brings up a browsing dialog box that is used to locate the directory of your choice. Consult the MATLAB manuals for setting the current directory in versions 5.2 or 5.3. Make sure that the current directory lists the directory (folder) that contains the M-files you want to run. "But why?" you might ask. When the name of a M-file is entered in the Command Window, the MATLAB system first looks in the current directory for the file to be interpreted. If not found there, it searches through directories on the search path and eventually checks its own file collection. An error message appears if the file can't be found, and this accounts for most of the problems new users of MATLAB experience when working with M-files. Setting the current directory properly ensures that your files are seen by the system. To make sure the directory is set properly, type `dir` in the Command Window and see if the files you are working with appear in the list. If not, go back and change the directory before proceeding.

The M-files for this manual were written using versions 5.2.1, 5.3, 6.0, and 6.1 of MATLAB. Both Macintosh and Windows platforms were used. A serious attempt was made to ensure they run with any of the versions mentioned, and this caused some awkward coding strategies. Every M-file can be downloaded from http://www.contemplinalg.com/, and they are organized by chapter and by section within a chapter. They are an integral part of the manual. Each directory (folder) contains an M-file `comments.m` that lists the M-files and all supporting M-files for that folder. This causes some duplication, but it is the lesser of two evils. Some files can't be read. They have a suffix p rather than m, and `solvesys.p` or `formatA.p` come to mind. These files contain material that is not considered appropriate for a linear algebra course.

Mistakes are bound to occur in a project of this magnitude. I hope they are few in number and easily corrected. Enjoy your MATLAB experience.

Herman Gollwitzer

October 2002

CONTENTS

Chapter 1

Vectors

INTRODUCTION

Technology exercises in Chapter 1 are designed to acquaint you with several MATLAB commands for entering, manipulating, and displaying vectors. The representation of a vector can be casual when you are writing it down on paper. It must be precise when you work with MATLAB. Loosely speaking, a vector can be thought of as a list, a row vector (matrix with one row), or a column vector(matrix with one column). For example, x, y, and z are all representations of a vector.

$$x = [1,2,3], \ y = \begin{bmatrix} 1 \\ 2 \\ 3 \end{bmatrix}, \text{ and } z = (1,2,3)$$

The three representations all hold the same information, but they have different structures. The representations in x and y are different because they have different shapes. This distinction is crucial when working with sums or products in MATLAB. MATLAB does not support the list structure suggested by z for most computations, and vectors will always be understood to be of the row or column type. You should be familiar with the command window before proceeding. If necessary, check your local computing environment for getting started with MATLAB. Once you launch MATLAB and set the current directory to this chapter's files, the rest is transparent. Script files called M-files are used extensively in what follows, and it is important to remember to have the current directory for MATLAB set to the directory containing the files you are preparing to examine.

In what follows, each technology exercise is illustrated with MATLAB code. Some of the code fragments are taken from the command window and others are M-files that produce output in the command window. Code fragments that begin with >> are taken from the command window. The mark >> is a prompt symbol generally found in the command window. It may be different for your system. MATLAB code is presented in a Courier font such as `help format`. Commands you see in this chapter can be entered in the command window and executed so that you can verify the calculations being discussed. Most problems have associated M-files that automate the process so that entering commands in the command window is optional. Do not include the prompt symbol as part of the command.

EXERCISE SET 1.1
TECHNOLOGY EXERCISES

T1. (***Numbers and numerical operations***) Read how to enter integers, fractions, decimals, and irrational numbers such as π and $\sqrt{2}$ .Check your understanding of the procedures by converting π, $\sqrt{2}$, and 1/3 to decimal form with various numbers of decimal digits in the display. Read about the procedures for performing the operations of addition, subtraction, multiplication, division, raising numbers to powers, and extraction of roots. Experiment with numbers of your own choosing until you feel you have mastered the techniques.

The command window is used for entering data manually from the keyboard. Data is usually associated with a variable name and several examples of data entry and assignment are listed. The command `disp` is one way to display variables or the result of a computation in the command window. Integers, fractions, and decimals are entered just as you think about them.

Integers:

```
>> x=54321;
>> disp(x)
        54321

>> x

x =
```

```
        54321
```

Notice how empty rows are put in to separate the command from the output or the next command. This is called the loose display format. The extra space can be eliminated in subsequent command window output with the command `format compact`. Repeating the commands shows the effect. The command `format loose` is the default display format.

```
>> format compact
>> x = 54321;
>> disp(x)
        54321
>> x
x =
        54321
```

Comments can be inserted after a command using the % symbol. Text that appears after this mark is treated as a comment and is not executed. This style is not recommended generally for computations in the command window because of the extra work involved. It is used in this manual to help explain commands and other features in a MATLAB context. Comments are optional on your part. A commented version of the last assignment is listed next.

```
>> x = 54321; % Assign an integer to x
>> disp(x) % Use disp for a neater looking format for x
        54321
>> x % or omit the semi-colon before pressing the return
x =
        54321
```

Fractions:

```
>> y = 355/113; % an approximation to π
>> disp(y) % current format is short and five digits are shown
     3.1416
>> format rat % request a rational format for subsequent display
>> disp(y) % agrees with the input because the numbers are reasonable
     355/113
>> format short % revert back to default decimal display mode
```

Decimals:

```
>> z = 43.214;
```

```
>> disp(z)   % Short format displays 5 digits, hence the 0
   43.2140
```

An approximation to π:

```
>> w = pi;
>> disp(w)  % a five decimal approximation is shown
   3.1416
```

The name `pi` stands for a built-in approximation to π that is accurate to about sixteen decimal places. A more accurate picture of `pi` is given below.

```
>> format long  % use sixteen decimals in display
>> format compact
>> pi
ans =
   3.14159265358979
>> format short
```

A numerical approximation to $\sqrt{2} = 2^{1/2}$:

```
>> g = 2^.5    % Omitting the semi-colon echoes the input immediately
g =
    1.4142
>> format long    % Ask for a longer display format
>> g
g =
   1.41421356237310
>> format short    % Select the default display format
```

Exponentiation is carried out using the symbol `^`. The format for decimal display can be controlled using the format command. Typing `help format` in the command window shows a description of the format command that is not listed here because of its length.

T2. (***Drawing vectors***) Read how to draw line segments in 2 or 3-dimensional space, and draw some line segments with initial and terminal points of your choice. If your utility allows you to create arrowheads, then you can make your line segments look like geometric vectors.

Drawing vectors interactively in MATLAB requires a figure window and scaling associated with that figure. The command `figure` causes a figure window to appear,

```
```

and the second command specifies scaling to provide a view of a box with extents $-3 \le x \le 3$, $-4 \le y \le 4$. Scaling limits are set with the vector $[-3,3,-4,4]$ in the axis command. The association $[-3,3,-4,4] \leftrightarrow [\text{left, right, bottom, top}]$ indicates how scaling is set for figures in general.

```
>> figure
>> axis([-3,3,-4,4])
```

The toolbar at the top of the figure window has several icons. Tools you will use are the diagonal line and the arrow that points in a northeasterly direction. Use the mouse and select the line icon. Place the mouse on the figure window, click and drag until you are satisfied with the line. Release the mouse and the line is finished. Placing lines in a figure window in this fashion has some limitations in that you can't resize or zoom the window later and expect the lines to adjust accordingly. Be sure to resize the window to your specifications before starting the line and arrow drawing process. Clicking on the arrow icon activates the arrow tool, and the procedure is identical to the line drawing process.

T3. (*Operations on Vectors*) Read how to enter vectors and how to calculate their sums, differences, and scalar multiples. Check your understanding of these operations by performing the calculations in Example 4.

Vectors are represented as row vectors or columns vectors. Everything is fine if you remain consistent in their usage.

```
>> x = [1, 2, 3]; % Separating commas are optional but recommended
>> disp(x)
     1     2     3

>> y = [1; 2; 3]; % A semi-colon separator creates a column vector
>> disp(y)
     1
     2
     3

>> z = [2 4 6]; %A row vector. Put at least one space between entries
>> disp(z)
     2     4     6
```

```
>> disp(x + y) % MATLAB protests if data shapes are mixed with + or -
??? Error using ==> +
Matrix dimensions must agree

>> disp(x+z) % vector addition
     3     6     9

>> disp(x - z) % vector subtraction
    -1    -2    -3

>> disp(2*x) % rescale x by 2
     2     4     6

>> disp(2*x - z) % Twice x minus z
     0     0     0
```

T4. Use your technology utility to compute the components of

$$\mathbf{u} = (7.1, -3) - 5\left(\sqrt{2}, 6\right) + 3(0, \pi)$$

to five decimal places.

The syntax for evaluating a linear combination of vectors follows the description given in the text. Be sure to include the multiplication operator *. MATLAB uses arithmetic that is accurate to about sixteen decimal places.

```
>> u = [7.1,-3]-5*[2^.5,6]+3*[0,pi];
>> disp(u)
    0.0289   -23.5752
```

EXERCISE SET 1.2
TECHNOLOGY EXERCISES

T1. **(Dot product and norm)** Some linear algebra programs provide commands for calculating dot products and norms, while others only provide a command for the dot product. In the latter case, norms can be computed from the formula $\|\mathbf{v}\| = \sqrt{\mathbf{v} \cdot \mathbf{v}}$.

Determine how to compute dot products and norms with your technology utility and perform the calculations in Examples 1, 2, and 4 of Section 2.1..

MATLAB provides commands for the dot product dot (x,y) and norm norm(x) for vectors x and y. The norm of the vector $\mathbf{v} = (-3,2,1)$ in Example 1 is approximately 3.7417 because

```
>> v = [-3,2,1];
>> disp(norm(v)) % calculate and display the norm of y
    3.7417
```

To find a unit vector **u** in the same direction as the vector $\mathbf{v} = (2,2,-1)$ in Example 2, simply rescale **v** by the reciprocal of its norm

```
>> v = [2,2,-1]; % row vector
>> u = (1/norm(v))*v; % include the extra parentheses for clarity
>> disp(u)
    0.6667     0.6667    -0.3333
```

A rational display format for u reproduces the result in the text.

```
>> format rat % request a rational display format
>> disp(u)
    2/3            2/3           -1/3
>> format short
>> disp(norm(u)) % quick check to verify that u is a unit vector
    1
```

The rational display format is something that is used frequently in this manual because many of the problems are phrased in terms of small integers or quotients of small integers. Answers for small problems are sometimes easier to interpret if such a format is used. There is an informal principle that characterizes computation and presentation in MATLAB:

What You See Is Usually Not What You Have.

The next code fragments all deal with the number 1/3 in a MATLAB environment. The literal 1/3 is converted to an internal binary format on the first line. Binary representations are unfamiliar to most people, and so a decision is made as to how x is displayed as a character string in the command window. The first display uses the default format called short. The next command asks for a rational format for subsequent output

7

and software converts the internal representation to a familiar form 1/3 because x is moderate in size. The command `format long` causes subsequent output to have a larger number of digits in a display. Each of the display formats gives a representation of x and only the second appears accurate. In reality, the internal representation for 1/3 is only accurate to about sixteen decimals. This marks a major distinction between MATLAB and a computer algebra system. This issue comes up repeatedly in what follows, and you should have no difficulty after you read the M-files that illustrate the different styles of output.

```
>> x = 1/3;
>> disp(x)
     0.3333

>> format rat
>> disp(x)
    1/3

>> format long
>> disp(x)
    0.33333333333333
```

Several other styles are discussed briefly when `help format` is typed in and executed.

T2. **(Sigma notation)** Determine how to evaluate expressions involving sigma notation and compute

$$\text{(a)} \quad \sum_{k=1}^{10} k^3 \quad \text{(b)} \quad \sum_{k=1}^{20} k^2 \cos(k\pi)$$

Keep in mind that finite sums can be interpreted as a dot product of suitable vectors. The sum in (a) is found in `val`.

```
>> x = 1:10; % x is the row vector [1 2,..,10]
>> y = x.^2; % square individual elements of x to get y
>> val = dot(x,y);
>> disp(val)
        3025
```

The sum can be interpreted as a dot product of a vector of ones with the vector of the cubes of the integers from 1 to 10 because $\sum_{k=1}^{10} k^3 = \sum_{k=1}^{10} 1 \times k^3$. This translates into MATLAB code that also evaluates the sum.

```
>> disp(dot(ones(1,10),(1:10).^3));
        3025
```

How does it work? The command `1:10` creates a row vector of integers from 1 to 10 that is a simple arithmetic progression. The command `(1:10).^3` distributes the cube operator onto each element of the row vector `(1:10)`. The steps are shown below so that you can review the results of each command.

```
>> disp(1:10)
     1     2     3     4     5     6     7     8     9    10
>> disp((1:10).^3)

     1     8    27    64   125   216   343   512   729  1000
```

Requesting `ones(1,10)` generates a vector of ones having 1 row and 10 columns

```
>> disp(ones(1,10))
     1     1     1     1     1     1     1     1     1     1
```

Try `ones(10,1)` to see a column vector of ones. Another strategy is to generate the cubes of the integers $1,2,\cdots,10$ and then add them up. This is the same as the last approach, and it avoids a direct appeal to the dot product.

```
>> val = sum((1:10).^3);
>> disp(val)
        3025
```

The command `(1:10).^3` generates the cubes of the integers and the `sum` command adds up the entries in this vector to give 3025. The period before `^` is important because MATLAB interprets the operator differently without it.

```
>> (1:10)^3
??? Error using ==> ^
Matrix must be square.
```

The sum in (b) follows the same pattern and we use a vector of squares k^2 and a vector of cosines $\cos(k\pi)$. The result is 210.

```
>> arg = 1:20; % vector of integers as an argument
>> x = arg.^2; % vector of squares of integers
>> y = cos(pi*arg); % vector of cosines at kπ
>> disp(dot(x,y)) % evaluate the sum
   210
```

It is instructive to verify a computation whenever possible. The expression $\cos(k\pi)$ alternates between 1 and –1 according as k is even or odd. In fact, $\cos(k\pi)=(-1)^k$ and the sum in (b) can be evaluated another way. The expression $(-1).\text{^arg})$ generates a vector of length 20 that consists of repetitions of the pattern $[1,-1]$. In summary,

```
>> disp(dot(x,(-1).^arg))
   210
```

T3. (a) Find the sine and cosine of the angle between the vectors **u** = (1, –2, 4, 1) and

   v = (7, 4, –3, 2).

The cosine is approximately –0.2655, and the sine is 0.9641.

```
>> u = [1,-2,4,1];
>> v = [7,4,-3,2];
>> c = dot(u,v)/(norm(u)*norm(v)); % cosine of angle between u and v
>> format rat % use a rational display format
>> disp(c)
  -252/949   % a rational format isn't always helpful
>> format short
>> disp(c)
   -0.2655 % the angle is in second quadrant
>> s = (1 - c^2)^.5; % use trigonometry to get the sine
>> disp(s)
    0.9641
>> disp(s^2+c^2) % verify that  sin²θ + cos²θ = 1
    1
```

(b) Find the angle between the vectors in part (a).

The MATLAB function acos calculates the arccosine and returns a radian angle in the interval $[0,\pi]$. The usual conversion factor gives the degree equivalent.

10

```
>> angle = acos(c); % inverse cosine
>> disp(angle) % in radians
    1.8396
>> deg_angle = angle*180/pi; % angle in degrees
>> disp(deg_angle)
  105.3992
```

T4. Use the method of Example 5 to estimate, to the nearest degree, the angles that a diagonal of a box with dimensions 10cm x 15cm x 25cm makes with edges of a box.

Associate x with 10cm, y with 15cm, and z with 25cm, and imagine a box similar to the one in Figure 1.2.6. A diagonal from the origin is represented in vector form as $d = (10,15,25)$. Vectors along the edges are $v_1 = (10,0,0)$, $v_2 = (0,15,0)$, and $v_3 = (0,0,25)$. Use the ideas in T3 and calculate the angles requested.

```
>> d = [10,15,25];
>> v1 = [10,0,0];
>> v2 = [0,15,0];
>> v3 = [0,0,25];
>> dnorm = norm(d);
>> anglex = acos(dot(d,v1)/(dnorm*norm(v1)));
>> anglex = acos(dot(d,v1)/(dnorm*norm(v1)));
>> disp(round(anglex*180/pi))
    71

>> angley = acos(dot(d,v2)/(dnorm*norm(v2)));
>> disp(round(angley*180/pi))
    61

>> anglez = acos(dot(d,v3)/(dnorm*norm(v3)));
>> disp(round(anglez*180/pi))
    36
```

The code `round(anglex*180/pi)` evaluates the angle and the rounds it off to the nearest integer. This example will test your patience because typing errors must be corrected as you proceed. When one occurs, hit the up arrow on your keyboard and the last command will appear. Edit it and proceed. A preferred way to enter several commands is to create a text or script file called an M-file that contains the commands. The MATLAB editor can be used. Go to the tool bar at the top of the command window and select the white icon on the left. This opens an empty text file that can be saved under a name of your choice in your current directory. The following M-file was created with the built-in editor in MATLAB and saved to a working directory with the name T4_1_2.

```
%T4_1_2
%This M-file carries out the calculations necessary
%for problem T4 in section 1.2. It relives the user
%from having to type in all the statements in the
%command window.

%echo on
d = [10,15,25]; % diagonal vector
v1 = [10,0,0]; % component along x-axis
v2 = [0,15,0]; % component along y-axis
v3 = [0,0,25]; % component along z-axis
dnorm = norm(d); % compute once and use 3 times
anglex = acos(dot(d,v1)/(dnorm*norm(v1))); % angle between d and v1
disp('The angle(degrees) between d and the x-axis is');
disp(round(anglex*180/pi));
%A cleaner way to display numbers in a phrase is described later.

angley = acos(dot(d,v2)/(dnorm*norm(v2)));
disp('The angle(degrees) between d and the y-axis is');
disp(round(angley*180/pi))

anglez = acos(dot(d,v3)/(dnorm*norm(v3)));
disp('The angle(degrees) between d and the z-axis is');
disp(round(anglez*180/pi))
%echo
```

The name 'T4_1_2' was typed in the command window as shown below. The lines below it are the command window outputs prescribed in the M-file.

```
>> T4_1_2
The angle(degrees) between d and the x-axis is
    71

The angle(degrees) between d and the y-axis is
    61

The angle(degrees) between d and the z-axis is
    36
```

You can also watch the text of an M-file as it executes. Remove the comment marks for echo on and echo off in this M-file, save it, and run it again. This time you will see the lines of the M-file along with the output. It is not shown here and the use of the echo command is optional. Be sure to put the comment mark % back if you don't want to see the M-file commands the next time you execute this M-file.

12

Using an M-file to organize computations has several merits. First, it forces you to think about what you want to compute with MATLAB. Second, any typing errors causes the M-file execution to cease and an error message is posted in the command window indicating where the error was detected. Simply edit to correct the error, save the M-file and start over. Third, comments you add to the M-file serve as documentation in case you have to explain it to someone or review it yourself. In summary, commented M-files are a crucial part of computing with MATLAB.

T5. (Sigma notation) Let u be the vector in R^{100} whose ith component is i, and let v be the vector in R^{100} whose ith component is $1/(i+1)$. Evaluate the dot product $u \bullet v$ by first writing it in sigma notation.

The dot product $u \bullet v$ is $\sum_{i=1}^{100} i/(i+1)$ and the sum is approximately 99.8027.

```
>> u = 1:100;
>> v = 1./(1+u); % note the use of the ./ operator
>> disp(dot(u,v))
   95.8027
```

The fraction 1./(1+u) is formed by first adding 1 to each element of u and then taking the reciprocal of each element. The result is a row vector of the same size. On a smaller scale,

$$1./[1,2,3] \Leftrightarrow [1,1/2,1/3]$$

The numerator is the scalar 1 and MATLAB interprets this as a row vector of ones for purposes of evaluation.

EXERCISE SET 1.3
TECHNOLOGY EXERCISES

T1. (*Parametric lines*) Many graphing utilities can graph parametric curves. If you have such a utility, then determine how to do this, and generate the $x = 5 + 5t$, $y = -7t$ (see Figure 8 in the text).

The M-file T1_1_3 gives the steps in creating the line.

```
%T1_1_3
% Purpose: Graph a parametric line
% x = 5 + 5t, y = -7t
figure % display a figure window
t = linspace(-3,3); % vector of 100 values over [-3,3].
x = 5 + 5*t; % parametric x-values
y = -7*t; % parametric y-values
plot(x,y); % plot points on the line for -3 <= t <= 3
%grid on % display a grid in the figure
```

Use the toolbar at the top of the figure to place text in that window. Click on the letter 'A' in the toolbar and then use the mouse to find a place for the text. Click the mouse and type in your text. Repeat as often as desired. The command linspace(-3,3) creates 100 equally spaced values between −3 and 3 for plotting purposes. More details on linspace are found by executing the command help linspace or doc linspace in the command window. A grid can be superimposed on the graph by deleting the comment mark % in the last line of the M-file and repeating the process. The keyboard command grid off removes the grid.

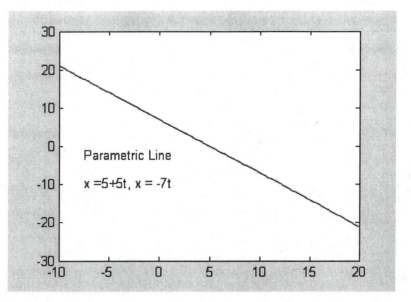

Parametric Line

x = 5+5t, x = -7t

T2. Generate the line L through the point (1,2) that is parallel to $v = (1,1)$. In the same window, generate the line through the point (1,2) that is perpendicular to L. If your lines do not look perpendicular, explain why.

The strategy used here is to create a figure window and set the scaling to include the rectangle $0 \le x \le 3$, $0 \le y \le 3$. After that, use the arrow tool in the tool bar to create the vector v, and the line tool to create the lines. The next MATLAB commands show how to create the figure window. Lines may not look perpendicular if the figure window is not sized manually to be square before starting this exercise.

```
>> figure % open a new figure window
>> axis([0,3,0,3]); % set the axis limits
```

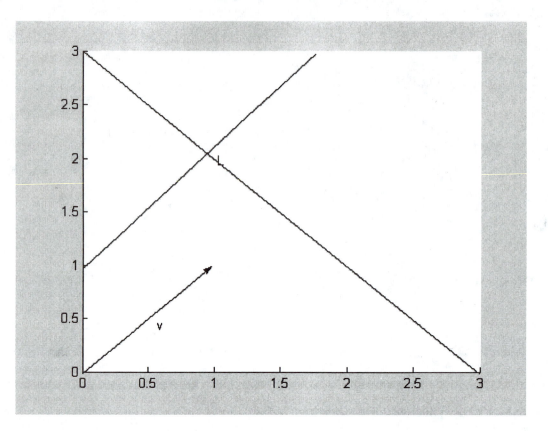

T3. Two intersecting planes in 3-space determine two angles of intersection, an acute angle $(0 \leq \theta \leq 90°)$ and its supplement $180° - \theta$ (see accompanying figure in the text). If $\mathbf{n}_1$ and $\mathbf{n}_2$ are normals to the planes, then the angle between $\mathbf{n}_1$ and $\mathbf{n}_2$ is θ or $180° - \theta$ depending on the directions of the normals. In each part, find the acute angle of intersection of the planes to the nearest degree.

(a) $x = 0$ and $2x - y + z - 4 = 0$

(b) $x + 2y - 2z = 5$ and $6x - 3y + 2z = 8$

The first pair of planes is $x = 0$ and $2x - y + z - 4 = 0$. Vectors normal to them are $\mathbf{n}_1 = (1,0,0)$ and $\mathbf{n}_2 = (2,-1,1)$, respectively, and the angle between the planes is calculated using the M-file T3a_1_3. The command window output gives the acute angle.

```
>> T3a_1_3
```

16

The acute angle between the planes is
     35

The M-file is listed here so that you can see how the MATLAB code mirrors the
mathematical steps when you do it by hand.

```
%T3a_1_3
% Find the acute angle between the planes
% whose normal vectors are n1 and n2.
n1 = [1 0 0]; % vectors normal to planes
n2 = [2,-1,1];
c = dot(n1,n2)/(norm(n1)*norm(n2)); % cosine of angle
angle = (180/pi)*acos(c); %in degrees
angle = round(angle); % to nearest degree
disp('The acute angle between the planes is')
disp(angle) % no need to subtract from 180 in this case
```

The second pair of planes is $x + 2y - 2z = 5$ and $6x - 3y + 2z = 8$. Vectors normal to the

planes are $n_1 = (1,2,-2)$ and $n_2 = (6,-3,2)$, respectively, and the angle between them is

calculated using the M-file T3b_1_3 which is not listed here.

```
>> T3b_1_3
The angle between the planes is
   101
The acute angle between the planes is
    79
```

T4. Find the acute angle of intersection between the plane $x - y - 3z = 5$ and the line

$$x = 2-t, \; y = 2t, \; z = 3t - 1$$

The acute angle of intersection between the plane $x - y - 3z = 5$ and the line $x = 2 - t$,

$y = 2t$, $z = 3t - 1$ is found by first calculating the angle $\theta \in [0,180]$ between a vector $\mathbf{n}$

normal to the plane and a vector $\mathbf{v}$ along the line. An acute angle between the plane and

the line is determined by subtracting θ from 90. The vector $\mathbf{n} = (1,-1,-3)$ is

perpendicular to the plane and $\mathbf{v} = (-1,2,3)$ is a vector along the line. The acute angle is

about 28.9 degrees and the command window results explain why.

```
>> n = [1,-1,3]; % vector normal to plane
```

```
>> v = [-1,2,3]; % vector along the line
>> cos_theta = dot(n,v)/(norm(n)*norm(v));
>> theta = (180/pi)*acos(cos_theta); % angle in degrees
>> angle = 90 - theta;
>> disp(angle)
   28.9138
```

The inverse cosine function `acos` returns an angle in the range $[0,\pi]$ and the conversion factor $180/\pi$ converts radians to degrees. Other inverse trigonometric functions such as `asin` (arcsine) or `atan` (aractan) also give angles in radians. These calculations are repeated in the M-file T4_1_3.

Summary

The technology exercises presented several MATLAB commands and problem solving styles you can use in subsequent chapters. An important aspect of your work with MATLAB is to understand the problem well enough so that you can create the correct data and then apply the appropriate data manipulation commands. Script files called M-files are a crucial part of working with MATLAB, and many of the computations were carried out using an M-file. Listed below are the commands introduced in this chapter. Brief descriptions were given as they were introduced, and additional details can be found in the MATLAB help files accessed from the tool bar at the upper right, or by using the command line help request. The latter requires that you know the name of the command. Type `help linspace` in the command window to get details on `linspace`. Type `help help` in the command window to get details on using help. Another source of information is in the documentation. Type `doc linspace` for a different explanation of `linspace`. The same strategy works for any other MATLAB command.

MATLAB Command Summary

MALAB commands introduced in this chapter are listed below for reference.

axis ----- axis([left,right,bottom,top]) sets the viewing axes limits for the current figure

cos ----- cos(x) gives the cosine of x

acos ----- acos(x) gives the inverse cosine of x in radians

% ----- text after % is treated as a comment and is not executed

disp ----- disp(x) displays a formatted version of x

dot ----- dot(x,y) evaluates the dot product of vectors x and y

echo ----- displays M-file line along with output in the command window. See help echo.

figure ----- creates a new figure window for plotting purposes

format short ----- default display format of five decimals

format rat ----- rational approximation format if the numbers are not too large or small

format long ----- display format with sixteen decimals.

help ----- command name for command line help

linspace ----- linspace(a,b) creates 100 equally spaced values between a and b

norm ----- norm(x) calculates the norm of the vector x

ones ----- ones(m,n) creates a matrix of ones with m rows and n columns

pi ----- an approximation to π that is part of MATLAB

plot ----- plot(x,y) plots and connects the dots with line segments for vectors x and y

sum ----- sum(x) adds up the entries in the vector x

round ----- round(x) rounds off x to the nearest integer

Chapter 2

Systems of Linear Equations

EXERCISE SET 2.1
TECHNOLOGY EXERCISES

T1. (*Linear Systems with Unique Solutions*) Solve the system in Example 6.

The system in question is

$$x + y + 2z = 9$$
$$2x + 4y - 3z = 1$$
$$3x + 6y - 5z = 0$$

There are several ways to solve it. The first uses a Symbolic Toolbox command `solve` that is available if you are using the Student Version of MATLAB. The command `rref` is available when that toolbox is not installed, and it mimics the gaussian elimination process used when working with paper and pencil.

To use the command `solve`, you must first create character versions of the equations to be solved. Single quotes surrounding the symbolic form of an equation establishes what is called a character string. The `solve` command uses this information to interpret the system and returns the results. Illustrated below are the steps needed to solve the system. Note that the output of solve recognizes the fact that there are three components to the solution which are x = 1, y = 2, and z = 3.

```
>> eqn1 = 'x+y+2*z=9'; % character version of first equation
>> eqn2 = '2*x+4*y-3*z=1'; % note the use of * for multiplication
>> eqn3 = '3*x+6*y-5*z=0';
>> [x,y,z] = solve(eqn1,eqn2,eqn3); % solution is on left side of =
>> format compact
>> disp(x)
1
>> disp(y)
2
>> disp(z)
3
```

Another approach uses the command `rref` that row reduces a matrix for subsequent interpretation. The first step is to create the augmented matrix as taken from the system in Example 6 and listed above. The row reduced matrix is determined and the solution is read off the result.

```
>> A = [1,1,2,9;2,4,-3,1;3,6,-5,0]; % create the augmented matrix
>> disp(A) % display augmented matrix A
     1     1     2     9
     2     4    -3     1
     3     6    -5     0
>> R = rref(A); % row reduce A
>> disp(R) % display it
     1     0     0     1
     0     1     0     2
     0     0     1     3
```

The row reduced form R is interpreted as an equivalent linear system. Row one is interpreted as $1x + 0y + 0z = 1$, or $x = 1$. The two remaining rows are interpreted in the same fashion and so $y = 2$ and $z = 3$. This approach is closer to what you carry out by hand, and it requires the most intervention on your part to interpret the results.

The last approach uses a utility function `solvesys` that is included in the files as part of the manual. A call to `solvesys` requires two inputs: the coefficient matrix A, and the right hand side b for a system described loosely as Ax = b. Using the system as a guide, fashion A from the coefficients of the variables on the left side.

$$x + y + 2z = 9$$
$$2x + 4y - 3z = 1$$
$$3x + 6y - 5z = 0$$

Be sure to include the signs of the coefficients.

$$A = \begin{bmatrix} 1 & 1 & 2 \\ 2 & 4 & -3 \\ 3 & 6 & -5 \end{bmatrix} \text{ and } b = \begin{bmatrix} 9 \\ 1 \\ 0 \end{bmatrix}$$

The three commands below indicate the steps in MATLAB.

```
>> A = [1 1 2;2 4 -3;3 6 -5];
>> b = [9;1;0];
>> solvesys(A,b)  % A first followed by b
*** *** *** *** *** *** ***
The system Ax = b is
|1   1   2|     |9|
|2   4  -3|x =  |1|
|3   6  -5|     |0|
*** *** *** *** *** *** ***
and the computed solution is
|x1|     |1|
|x2|  =  |2|
|x3|     |3|
*** *** *** *** *** *** ***
```

The output consists of a running commentary on the problem you submitted and an interpretation of the linear system described by the two inputs. Think of `solvesys` as a report generator that accepts A and b and displays results in the command window. The results are not available for use by other MATLAB commands. There are some restrictions on the inputs. There can be at most 5 linear equations in 8 unknowns. Thus the matrix *A* can have no more than five rows and eight columns. The matrices A and b can be symbolic matrices in case the Symbolic Toolbox is installed. Data for the report is generated using `rref`. More is said later concerning `solvesys`.

T2. Use your technology utility to determine whether the vector $\mathbf{u} = (8, -1, -7, 4, 0)$ is a linear combination of the vectors $(1, -1, 0, 1, 3)$, $(2, 3, -2, 2, 1)$, and $(-4, 2, 5, 0, 7)$, and if so express $\mathbf{u}$ as such a linear combination.

The problem is to decide if the vector $u = (8, -1, -7, 4, 0)$ is a linear combination of the three fixed vectors

$$a_1 = (1, -1, 0, 1, 3), \ a_2 = (2, 3, -2, 2, 1), \ a_3 = (-4, 2, 5, 0, 7)$$

In vector terms, ask if the system $x_1 a_1 + x_2 a_2 + x_3 a_3 = u$ has any solution? One strategy is to recast the question so that it refers to a linear system. Look closely at the system

$x_1a_1 + x_2a_2 + x_3a_3 = u$. It is written vertically so that you can see the pattern. Portrayed on the left is the symbolic linear combination, and its explicit form is on the right.

$$x_1a_1 = x_1\left(1,-1,0,1,3\right)$$
$$+ \qquad\qquad +$$
$$x_2a_2 = x_2\left(2,3,-2,2,1\right)$$
$$+ \qquad\qquad +$$
$$x_3a_3 = x_3\left(-4,2,5,0,7\right)$$
$$\updownarrow$$
$$u = \left(8,-1,-7,4,0\right)$$

Two vectors are equal when all their coordinates match. Focus on the first coordinate of the vectors on the right. It translates into the scalar equation

$$x_1 + 2x_2 - 4x_3 = 8$$

The remaining coordinates must match up too. There are four more equations and the entire list is

$$x_1 + 2x_2 - 4x_3 = 8$$
$$-x_1 + 3x_2 + 2x_3 = -1$$
$$0x_1 - 2x_2 + 5x_3 = -7$$
$$x_1 + 2x_2 + 0x_3 = 4$$
$$3x_1 + x_2 + 7x_3 = 0$$

It is easy to remember how this works. Note that the coefficients of x_1 are precisely the components of a_1. Similar results hold for the other two variables. The vector **u** is found on the right side of the equality, and this observation lets us set up and solve the system. The MATLAB code below carries out these steps and displays the row reduced form R of the system. The expression `[a1,a2,a3,u]` stacks the four column vectors horizontally and forms the 5x4 matrix that is the input matrix for `rref`.

$$[a_1, a_2, a_3, u] \leftrightarrow \begin{bmatrix} 1 & 2 & -4 & 8 \\ -1 & 3 & 2 & -1 \\ 0 & -2 & 5 & -7 \\ 1 & 2 & 0 & 4 \\ 3 & 1 & 7 & 0 \end{bmatrix}$$

```
>> a1=[1;-1;0;1;3]; % column vector for first vector
>> a2 = [2;3;-2;2;1];
>> a3=[-4;2;5;0;7];
>> u=[8;-1;-7;4;0]; % column vector for u
>> R = rref([a1,a2,a3,u]); % row reduce and interpret visually
>> disp(R)
     1     0     0     2
     0     1     0     1
     0     0     1    -1
     0     0     0     0
     0     0     0     0
```

Interpret R to mean that the original system $x_1 a_1 + x_2 a_2 + x_3 a_3 = \mathbf{u}$ can be solved using 2, 1, and -1 for x_1, x_2, and x_3, respectively. The values aren't really needed to answer the question in the affirmative, but they offer convincing evidence. It would be better if these commands had been put in an M-file in case further computations are needed. A coomputed solution can be checked by hand, but it is possible to let MATLAB verify the fact that the non-zero numbers in the last column of R are the weights x_1, x_2, and x_3. The zero vector should appear after $x_1 a_1 + x_2 a_2 + x_3 a_3 - \mathbf{u}$ is evaluated, and the code below confirms it.

```
>> x = R(1:3,4); % extract the first 3 entries from column 4
>> disp(x(1)*a1+x(2)*a2+x(3)*a3 - u) % form the linear combination
     0
     0
     0
     0
     0
```

The report generator `solvesys` confirms the results demonstrated earlier.

```
>> solvesys([a1,a2,a3],u)
*** *** *** *** *** *** ***
The system Ax = b is
| 1   2  -4|    | 8|
|-1   3   2|    |-1|
```

25

```
| 0  -2   5|x =  |-7|
| 1   2   0|      | 4|
| 3   1   7|      | 0|
*** *** *** *** *** *** ***
The computed solution is

|x1|    | 2|
|x2| = | 1|
|x3|    |-1|


*** *** *** *** *** *** ***
```

EXERCISE SET 2.2
TECHNOLOGY EXERCISES

T1. (**Reduced Row Echelon Form**) Read your documentation on how to enter matrices and how to produce reduced row echelon forms. Check your understanding of these commands by finding the reduced row echelon form of the matrix

$$\begin{bmatrix} 2 & -3 & 1 & 0 & 4 \\ 1 & 1 & 2 & 2 & 0 \\ 3 & 0 & -1 & 4 & 5 \\ 1 & 6 & 5 & 6 & -4 \end{bmatrix}$$

The process is to enter the matrix and use the `rref` command as shown below. Note how the matrix A was entered by rows. No separating commas were used to separate entries, but you must remember to put at least one space between the row values. It is safer to use commas to separate values when entering data.

```
>> A = [2 -3 1 0 4;1 1 2 2 0;3 0 -1 4 5;1 6 5 6 -4]; % matrix by rows
>> disp(A) % it is the correct matrix
     2    -3     1     0     4
     1     1     2     2     0
     3     0    -1     4     5
     1     6     5     6    -4

>> disp(rref(A)) % display the row reduced echelon form
   1.0000        0        0   1.3077   1.5000
        0   1.0000        0   0.8462  -0.5000
        0        0   1.0000  -0.0769  -0.5000
        0        0        0        0        0

>> format rat % use rational format because A has modest entries
>> disp(rref(A)) % what you get if you do it by hand and keep fractions
```

1	0	0	17/13	3/2
0	1	0	11/13	-1/2
0	0	1	-1/13	-1/2
0	0	0	0	0

Another version even returns the pivot columns seen in the row reduced form.

```
>> [R,pivotcols] = rref(R);
>> disp(R)
    1.0000         0         0    1.3077    1.5000
         0    1.0000         0    0.8462   -0.5000
         0         0    1.0000   -0.0769   -0.5000
         0         0         0         0         0

>> disp(pivotcols)
     1     2     3
```

This relieves us of manually scanning the row reduced form for pivot columns. When the matrix is symbolic, the same rref command carries out the usual operations.

```
>> syms a b % declare symbolic variables
>> A = [1, 2*a,3;b,1,-b^2]
A =

[    1,   2*a,     3]
[    b,     1, -b^2]
>> R = rref(A) % row reduce a symbolic matrix
R =

[1,      0, -(3+2*b^2*a)/(-1+2*b*a)]
[0,      1,        b*(b+3)/(-1+2*b*a)]
>> [R,pivots] = rref(A) % symbolic version doesn't give pivots
??? Error using ==> rref
Too many output arguments.
```

The MATLAB system protested because rref for symbolic matrices does not also give pivots. A user-defined function rrefsym is included with the M-files for this chapter, and it does optionally return pivot columns. It isn't important to know how it works at this time.

```
>> R = rrefsym(A) % with one output
R =

[1,      0, -(3+2*b^2*a)/(-1+2*b*a)]
[0,      1,        b*(b+3)/(-1+2*b*a)]

>> [R,pivots] = rrefsym(A) % returns pivot columns too
R =
```

```
[1,      0, -(3+2*b^2*a)/(-1+2*b*a)]
[0,      1,      b*(b+3)/(-1+2*b*a)]

pivots =

    1    2
```

A word on order in the outputs is in order. The first output name must always be associated with the row reduced echelon form. What follows is correct code but confusing code.

```
>> [pivots,R] = rrefsym(A)
pivots =

[1,      0, -(3+2*b^2*a)/(-1+2*b*a)]
[0,      1,      b*(b+3)/(-1+2*b*a)]

R =

    1    2
```

T2. (***Inconsistent linear systems***) Technology utilities will often successfully identify inconsistent linear systems, but they can sometimes be fooled into reporting an inconsistent system as consistent or vice versa. This typically occurs when some of the numbers that occur in the computations are so small that round off error makes it difficult for the utility to determine whether or not they are equal to zero. Create some inconsistent linear systems and see how your utility handles them.

It takes some thought to come up with meaningful examples. An augmented matrix for an inconsistent 2x2 system is

$$\begin{bmatrix} 10 & 20 & 2 \\ 10\varepsilon & 20\varepsilon & 3\varepsilon \end{bmatrix}$$

The lower right element forces the inconsistency. The symbol ε stands for a small, built-in number called eps in MATLAB, and its approximate value is 2.2204e-016. The output from the M-file T2_2_2 indicates that the row reduction process couldn't decide how to handle extremely small numbers properly, and it caused the last row to be reported as all zeros. A correct interpretation of the computed row reduced form is that the system has infinitely many solutions. It is a correct interpretation of incorrect data. The correct row reduced augmented matrix below leads to the conclusion that the system has no solution.

$$\begin{bmatrix} 1 & 2 & 0.2 \\ 0 & 0 & 1 \end{bmatrix}$$

```
>> T2_2_2
The augmented matrix for an inconsistent system
  1.0000e+001   2.0000e+001   2.0000e+000
  2.2204e-015   4.4409e-015   6.6613e-016


The row reduced form R suggests there are
infinitely many solutions. R is
  1.0000e+000   2.0000e+000   2.0000e-001
        0             0             0
```

T3. (*Linear systems with infinitely many solutions*) The commands used by technology utilities for solving linear systems all handle systems with infinitely many solutions differently. See what happens when you solve the system in Example 5 of this section.

The system in question is

$$\begin{bmatrix} 1 & 3 & -2 & 0 & 2 & 0 \\ 2 & 6 & -5 & -2 & 4 & -3 \\ 0 & 0 & 5 & 10 & 0 & 15 \\ 2 & 6 & 0 & 8 & 4 & 18 \end{bmatrix} x = \begin{bmatrix} 0 \\ -1 \\ 5 \\ 6 \end{bmatrix}$$

One strategy is to row reduce the associated augmented matrix and interpret the results. This requires human intervention and there is no command in MATLAB to solve a system that might have an infinite number of solutions. That is why solvesys is provided with the M-files for this chapter. Details are in the M-file T3_2_2.

```
>> T3_2_2
A is
     1          3         -2          0          2          0
     2          6         -5         -2          4         -3
```

```
         0          0          5         10          0         15
         2          6          0          8          4         18

b is
         0
        -1
         5
         5
The augmented matrix [A,b] is
         1          3         -2          0          2          0          0
         2          6         -5         -2          4         -3         -1
         0          0          5         10          0         15          5
         2          6          0          8          4         18          5

The corresponding row reduced form R of [A,b] is
         1          3          0          4          2          0          1
         0          0          1          2          0          0        1/2
         0          0          0          0          0          1        1/6
         0          0          0          0          0          0          0
```

The last row of zeros in the row reduced form signals that the system is consistent. The pivots occur in columns 1, 3, and 6, and hence x_2, x_4, x_5 are free variables. Each row can be interpreted as an equation for an equivalent system, and a solution representation is

$$x_2 = t_1 \leftrightarrow \text{free}$$
$$x_4 = t_2 \leftrightarrow \text{free}$$
$$x_5 = t_3 \leftrightarrow \text{free}$$
$$x_1 = -3t_1 - 4t_2 - 2t_3 + 1$$
$$x_3 = -2t_2 + \frac{1}{2}$$
$$x_6 = \frac{1}{6}$$

It is important to realize that there is no unique way to describe a solution when a linear system has infinitely many solutions. Display output from solvesys describes the infinitely many solutions as a linear combination of vectors and free variables.

```
>> solvesys(A,b)  % The inputs don't have to be named A and b
*** *** *** *** *** *** ***
The system Ax = b is
|1    3   -2    0    2    0| |x1|     | 0|
|2    6   -5   -2    4   -3| |x2|  =  |-1|
|0    0    5   10    0   15| |x3|     | 5|
|2    6    0    8    4   18| |x4|     | 5|
                            |x5|
                            |x6|
```

A computed solution is

```
|x1|    | 1|       |-3|       |-4|       |-2|
|x2|    | 0|       | 1|       | 0|       | 0|
|x3| = |1/2| + t1| 0| + t2|-2| + t3| 0|
|x4|    | 0|       | 0|       | 1|       | 0|
|x5|    | 0|       | 0|       | 0|       | 1|
|x6|   |1/6|       | 0|       | 0|       | 0|
```

The free parameters t1, t1, t2, t3 appear because
the system has infinitely many solutions.

You can see from this output that there is total agreement on the free variables and the solution form. This happens because `solvesys` uses `rref` and interprets the row reduced form in the same way you would if you did it by hand. When collapsed to a single vector, the solution reads

$$\begin{bmatrix} x_1 \\ x_2 \\ x_3 \\ x_4 \\ x_5 \\ x_6 \end{bmatrix} = \begin{bmatrix} -3t_1 - 4t_2 - 2t_3 \\ t_1 \\ -2t_2 \\ 1/4 + t_2 \\ t_3 \\ 1/6 \end{bmatrix}$$

T4. Solve the linear system

$$\tfrac{1}{5}x + \tfrac{1}{4}y + \tfrac{1}{2}z = \tfrac{37}{120}$$
$$\tfrac{1}{3}x + \tfrac{1}{7}y + \tfrac{1}{4}z = \tfrac{93}{336}$$
$$\tfrac{1}{4}x + \tfrac{1}{6}y + \tfrac{1}{3}z = \tfrac{43}{180}$$

The augmented matrix for this system is

$$\begin{bmatrix} 1/5 & 1/4 & 1/2 & 37/120 \\ 1/3 & 1/7 & 1/4 & 93/336 \\ 1/4 & 1/6 & 1/3 & 43/180 \end{bmatrix}$$

Data entry errors are common when creating such a matrix directly in the command window. A better strategy is to create an easily modified M-file that generates the matrix

31

and displays the row reduced echelon form so that the solution structure can be interpreted. That file is listed next, followed by the row reduced form that appeared in the command window.

```
% T4_2_2
% Commas are used to separate data items
% A is the coefficient matrix
A = [1/5,1/4,1/2
     1/3,1/7,1/4
     1/4,1/6,1/3];
b = [37/120; 93/336; 43/180]; % column vector
Ab = [A,b]; % augmented matrix
R = rref(Ab);

% display code
format rat % rational display
disp('R is');
disp(R);
format short % default display format
```

The command window output below is clear in context, but is best to echo the input data too so that you know what you working with. Doing so lets you catch data entry errors. Show A and b at the very least. Display formatting commands are inserted in the M-file to control the output format when results are displayed in the command window.

```
>> T4_2_2
The augmented matrix [A,b] is
     1/5          1/4          1/2          37/120
     1/3          1/7          1/4          31/112
     1/4          1/6          1/3          43/180


The row reduced form R is
     1            0            0            2/7
     0            1            0            47/15
     0            0            1          -149/140
```

The unique solution is found in the last column of R: $x = 2/7$, $y = 47/15$, and $z = -149/40$. Of course, solvesys can also be used to confirm the result. Roundoff errors are part of numerical computations, and they can have a small effect on the reported results. It doesn't seem to be the case here.

```
>> solvesys(A,b)
*** *** *** *** *** *** ***
The system Ax = b is
```

```
|1/5  1/4  1/2||x1|     |37/120|
|1/3  1/7  1/4||x2| =  |31/112|
|1/4  1/6  1/3||x3|     |43/180|
*** *** *** *** *** *** ***
```
The computed solution is

```
|x1|   |      2/7|
|x2| = |    47/15|
|x3|   |-149/140|
```

There may be roundoff errors in the computed solution.
```
*** *** *** *** *** *** ***
```

T5. In each part find values of the constants that make the equation an identity.

(a)
$$\frac{3x^3 + 4x^2 - 6x}{(x^2 + 2x + 2)(x^2 - 1)} = \frac{Ax + B}{x^2 + 2x + 2} + \frac{C}{x-1} + \frac{D}{x+1}$$

(b)
$$\frac{3x^4 + 4x^3 + 16x^2 + 20x + 9}{(x+2)(x^2+3)^2} = \frac{A}{x+2} + \frac{Bx + C}{x^2 + 3} + \frac{Dx + E}{\left(x^2 + 3\right)^2}$$

[*Hint*: Obtain a common denominator on the right, and then equate corresponding coefficients of the various powers of x in the two resulting numerators. Students of calculus may recognize this as a problem in partial fractions.]

There are two approaches for solving this problem and neither one follows the hint. One is symbolic and the other is strictly numeric. The numeric approach is taken up later. The patterns in the partial fraction expansion (a) indicate a linear relation involving A, B, C, and D. Substitute $x = 2$ into

$$\frac{3x^3 + 4x^2 - 6x}{(x^2 + 2x + 2)(x^2 - 1)} = \frac{Ax + B}{x^2 + 2x + 2} + \frac{C}{x-1} + \frac{D}{x+1}$$

to obtain

$$\frac{1}{5}A + \frac{1}{10}B + C + \frac{1}{3}D = \frac{28}{30}$$

33

Choosing values of x other than 1 and -1 provides three other linear equations. The choice is not important, and the resulting system of four equations in four unknowns can be solved to obtain values for A, B, C, and D. Be sure not to select values that cause any of the denominators to vanish. Command window output below describes the process and also shows that different substitution choices for x don't matter. MATLAB symbolic toolbox commands are used for manipulating the partial fraction expansion.

```
>> T5asym_2_2
The rational function equation is
(A*x+B)/(x^2+2*x+2)+C/(x-1)+D/(x+1)  =
     (3*x^3+4*x^2-6*x)/(x^2+2*x+2)*(x^2-1)
  ***    ***    ***    ***
Substitute x = -2, 0, 2, and -3 to obtain four equations
The four equations to be solved are:
-A+1/2*B-1/3*C-D = 2/3
1/2*B-C+D = 0
1/5*A+1/10*B+C+1/3*D = 14/15
-3/5*A+1/5*B-1/4*C-1/2*D = -27/40
  ***    ***    ***    ***
The solution is
A = 32/5
B = 36/5
C = 1/10
D = -7/2
****************************************
Repeat the process with different substitutions.
  ***    ***    ***    ***
Substitute x = 4, 0, 2, and 3 to obtain four equations.
The four equations to be solved are:
2/13*A+1/26*B+1/3*C+1/5*D = 116/195
1/2*B-C+D = 0
1/5*A+1/10*B+C+1/3*D = 14/15
3/17*A+1/17*B+1/2*C+1/4*D = 99/136
  ***    ***    ***    ***
The solution is the same as in the first case.
A = 32/5
B = 36/5
C = 1/10
D = -7/2
```

The second part of the problem deals with a similar partial fraction expansion. Find A, B, C, D, and E when

$$\frac{3x^4 + 4x^3 + 16x^2 + 20x + 9}{(x+2)(x^2+3)^2} = \frac{A}{x+2} + \frac{Bx+C}{(x^2+3)} + \frac{Dx+E}{(x^2+3)^2}$$

There are five unknowns and the previous strategy is repeated. The M-file T5bsym_2_2 documents the details, and it should be compared with T5asym_2_2 because different techniques are used when manipulating the data.

```
>> T5bsym_2_2
The rational function equation is
A/(x+2)+(B*x+C)/(x^2+3) + (D*x+E)/(x^2+3)^2 =
        (3*x^4+4*x^3+16*x^2+20*x+9)/((x+2)*((x^2+3)^2))
   ***    ***    ***    ***
Substitute x = -1:3 to obtain five equations
The five equations to be solved are:
A-1/4*B+1/4*C-1/16*D+1/16*E = 1/4
1/2*A+1/3*C+1/9*E = 1/2
1/3*A+1/4*B+1/4*C+1/16*D+1/16*E = 13/12
1/4*A+2/7*B+1/7*C+2/49*D+1/49*E = 193/196
1/5*A+1/4*B+1/12*C+1/48*D+1/144*E = 47/60
   ***    ***    ***    ***
The solution is
A = 1
B = 2
C = 0
D = 4
E = 0
*****************************************
Repeat the process with different substitutions.
   ***    ***    ***    ***
Substitute x = 0:4 to obtain five equations
The five equations to be solved are:
1/2*A+1/3*C+1/9*E = 1/2
1/3*A+1/4*B+1/4*C+1/16*D+1/16*E = 13/12
1/4*A+2/7*B+1/7*C+2/49*D+1/49*E = 193/196
1/5*A+1/4*B+1/12*C+1/48*D+1/144*E = 47/60
1/6*A+4/19*B+1/19*C+4/361*D+1/361*E = 1369/2166
   ***    ***    ***    ***
The solution is the same as in the first case.
A = 1
B = 2
C = 0
D = 4
E = 0
```

A numeric approach does not benefit from the symbolic substitution capabilities found in the Symbolic Toolbox. In the symbolic approach, the rational function expression was created and substitutions for x were made to obtain a linear system. For a numeric approach, substitutions are made into the polynomials that define the rational functions, and then the quotients are evaluated. The effect is the same, but only the coefficients of the unknowns are calculated. A quick look is provided after polynomial products are discussed.

MATLAB provides a function for multiplying two polynomials when they are represented in vector form. It is called `conv`, which is short for convolution. It happens that polynomial multiplication is just one example of convolution, and courses in applied mathematics or digital signal processing take up the issue of convolution. Treat it as a black box that calculates polynomial products. To illustrate the use of `conv`, consider two polynomials

$$[3,2,1] \Leftrightarrow 3x^2 + 2x + 1 \text{ and } [1,-1] \Leftrightarrow x - 1$$

Their product is $3x^3 - x^2 - x - 1$, and the MATLAB code below illustrates how to use `conv`. Polynomial evaluation works for vector arguments too. It is possible to evaluate the rational function

$$\frac{3x^2 + 2x + 1}{x - 1}$$

at several points without much effort. The next code fragments illustrate these points.

```
>> p=[3 2 1]; % p(x) = 3x^2 + 2x + 1
>> q = [1 -1]; % q(x) = x - 1
>> r = conv(p,q); % the product of p and q
>> disp(r) % vector of coefficients for product
     3     -1     -1     -1
>> s = conv(q,p); % convolution is a commutative operation
>> disp(s) % same polynomial
     3     -1     -1     -1
>> x = [-1;0]; % column of evaluation points
>> val = polyval(p,x)./polyval(q,x); % evaluate p(x)/q(x) at -1 and 0
>> disp(val)
    -1
    -1
```

An M-file `polymanip` reproduces these commands and the output.

The heart of the symbolic approach consists of being able to substitute into a rational function expression to obtain a linear system. With appropriate applications of `polyval` and `conv`, a numeric version of the linear system for A, B, C, and D can be formulated and solved using the linear template

$$\frac{3x^3 + 4x^2 - 6x}{(x^2 + 2x + 2)(x^2 - 1)} = \frac{x}{x^2 + 2x + 2}A + \frac{1}{x^2 + 2x + 2}B + \frac{1}{x-1}C + \frac{1}{x+1}D$$

This approach is documented in the M-file T5anum_2_2 and no output is given because the results are identical.

EXERCISE SET 2.3
TECHNOLOGY EXERCISES

Linear algebra ideas are found in many problems, and it is so easy to emphasize concepts rather than uses in an introductory course. It is more common that you see the consequences of linear algebra rather than the background details. A function `trigwheel` illustrates a trigonometric interpolation problem and builds a movie that plays several times. Solving a linear system many times is what it takes to produce the movie. I suggest you make the Command Window small and position a figure window in the middle of the screen before executing the M-file. The plotting techniques are not sophisticated and there is considerable flickering in the Command Window until the movie is complete. These noxious effects can be eliminated, but it requires more knowledge of MATLAB's extensive graphics system.

```
>> trigwheel % does it all
```

T1. Investigate your technology's commands for finding interpolating polynomials, and then confirm your understanding of these commands by checking the result obtained in Example 6.

Find the interpolating polynomial $p(x) = a_0 + a_1 x + a_2 x^2 + a_3 x^3$ for the data $(1,3)$, $(2,-2)$, $(3,-5)$, and $(4,0)$. There is a function called `polyfit` that makes it easy to determine the coefficients of an interpolating polynomial. Otherwise, you have to build a linear system based on the polynomial model and solve it to find the polynomial coefficients.

M-file T1_2_3 makes clear how `polyfit` is used. The third argument in the call below is crucial, and the help file for `polyfit` gives more details than you will probably want to see at this point. MATLAB does not provide a convenient way to describe a polynomial in traditional terms except when the coefficients are interpreted as symbolic variables. Enter the command

```
>> doc poly2sym
```

to see a full explanation. The function `poly2sym` converts the numeric coefficients into a traditional polynomial format, and it has some unexpected side effects.

```
>> p = poly2sym([pi exp(1)]);
>> disp(p)
pi*x+3060513257434037/1125899906842624
```

An approximation to exp(1) is first converted to a rational approximation before it is used as a coefficient. The function `polyrep` carries out the task of building a character representation of a polynomial when the coefficients are numeric.

```
>> p = polyrep([pi exp(1)]);
>> disp(p)
3.1416x + 2.7183
```

This expression is more familiar and `polyrep` is included with the M-files for this chapter. The M-file for this problem is listed in part so that you can see how the coefficients are determined.

```
% T1_2_3
format short
%Find a cubic polynomial through four points
x = [1;2;3;4]; % column vector
y = [3;-2;-5;0];
disp('Data for interpolating polynomial of degree 3')
disp('     x      y'); % header for table
disp([x,y]) % horizontal concatenation of x and y to form a table
n = length(x); % number of data points
% get coefficients p for interpolating polynomial
p = polyfit(x,y,n-1); % the third argument is crucial
% n-1 forces interpolation with a polynomial of
% degree (n-1) based on n points

% The remaining commands are for display purposes
```

The resulting output shows the data and the interpolating polynomial

$$p(x) = 4 + 3x - 5x^2 + x^3.$$

```
>> T1_2_3
Data for interpolating polynomial of degree 3
      x       y
      1       3
      2      -2
      3      -5
      4       0

polynomial coefficients, MATLAB style.
      1.0000    -5.0000    3.0000    4.0000

A more familiar representation using polyrep
p(x) = x^3 - 5x^2 + 3x + 4

The displayed polynomial coefficients are not quite
what they appear to be. Using format long e as a display
format, you see something a little different.
polynomial coefficients again:
      1.000000000000002e+000
     -5.000000000000012e+000
      3.000000000000019e+000
      3.999999999999989e+000
```

T2. MATLAB has several commands that make is easy to generate and plot polynomials.

   (a) Use your technology utility to find the polynomial of degree 5 that passes through the points

   $$(1, 1), (2, 3), (3, 5), (4, -2), (5, 11), (6, -12).$$

   (b) Follow the directions in part (a) for the points (1, 1), (2, 4), (3, 9), (4, 16), (5, 25), (6, 36). Give an explanation for what happens.

The interpolating polynomial through the five points $(1,1)$, $(2,3)$, $(3,5)$, $(4,-2)$, $(5,11)$, and $(6,-12)$ is found by using the function polyfit. This function constructs coefficients for an interpolating polynomial, and more. Type doc polyfit for an extensive explanation from the online help files. For now, we need only to understand how to use it for interpolation.

```
>> help polyfit
```

39

```
POLYFIT Fit polynomial to data.
   POLYFIT(X,Y,N) finds the coefficients of a polynomial P(X) of
   degree N that fits the data.
```

The subtle part is to choose the correct value of N. It takes 2 points to build a line and 3 points to get an interpolating quadratic, and so on. If n is the number of data points, then the call $p = $ `polyfit(x,y,n-1)` gives the coefficients of an interpolating polynomial of degree at most n-1. The x-values must be distinct. Using any other third argument leads to other data fitting situations that are taken up later in the manual when least squares data fitting is an issue. The code fragments indicate how it is used when finding a line $y = b + mx$ through the points (0,1) and (1,3).

```
>> x = [0 1];
>> y = [1 3];
>> n = 2; % number of data points
>> p = polyfit(x,y,n-1)
p =

      2      1
```

A polynomial p in MATLAB starts from the highest power down, and thus $m = 2$ and $b = 1$. Be sure to read the M-file T2a_2_3 for details.

```
>> T2a_2_3
x-values
     1      2      3      4      5      6
y-values
     1      3      5     -2     11    -12
Interpolating polynomial coefficients, MATLAB style
   -1.0250   16.9583 -104.4583   295.0417 -374.5167  169.0000
Polynomial evaluation at the interpolation points
verifies that the polynomial indeed interpolates.
Row 1 is x and row 2 is p(x)
    1.0000    2.0000    3.0000    4.0000    5.0000    6.0000
    1.0000    3.0000    5.0000   -2.0000   11.0000  -12.0000
```

Next, repeat the exercise from part (a) with the data $(1,1)$, $(2,4)$, $(3,9)$, $(4,16)$, $(5,25)$, and $(6,36)$.

```
>> T2b_2_3
x-values
     1      2      3      4      5      6
```

```
y-values
      1      4      9     16     25     36
Interpolating polynomial coefficients, MATLAB style
     0.0000    -0.0000     0.0000     1.0000     0.0000    -0.0000
Polynomial evaluation at the interpolation points
verifies that the polynomial indeed interpolates.
Row 1 is x and row 2 is the corresponding value p(x)
     1.0000     2.0000     3.0000     4.0000     5.0000     6.0000
     1.0000     4.0000     9.0000    16.0000    25.0000    36.0000
```

A graph of the data and the interpolating polynomial are part of the output. It is not shown here. The resulting polynomial is a simple quadratic that exactly fits the data. A closer inspection of the polynomial coefficients shows that the original data describes a quadratic function. The formatted coefficients include –0.0000. This is how a slightly negative number is sometimes displayed. The coefficients are called a in the M-file, and the next commands provide additional insight.

```
>> format long
>> disp(a') % display row vector a in column form

   0.00000000000000
  -0.00000000000001
   0.00000000000003
   0.99999999999991
   0.00000000000011
  -0.00000000000004
```

Roundoff errors in the numeric computations cause the computed coefficients to be off a little in the fourteenth decimal place.

T3. Find integer values of the coefficients for which the equation
$$A(x^2 + y^2 + z^2) + Bx + Cy + Dz + E = 0$$
is satisfied by all of the following ordered triples (x, y, z):
$$(\tfrac{1}{2}, -\tfrac{1}{3}, 1), \quad (\tfrac{1}{3}, -\tfrac{1}{2}, \tfrac{2}{3}), \quad (2, -\tfrac{1}{3}, \tfrac{1}{2}), \quad (1, 0, 1)$$

Note that the form of the equation leads to a homogeneous system for A, B, C, D, and E. There is no unique solution. A strategy is to substitute the coordinates into the equation and create a homogeneous linear system with 4 equations and 5 unknowns. Thinking ahead, any MATLAB code should be organized along the following lines.

41

1. Create the points as row vectors.

2. Substitute the coordinates into the equation. Each substitution leads to another equation.

3. Form and row reduce the matrix for the resulting homogeneous system.

4. Interpret the results and find an integer solution for the unknowns.

This strategy is followed in the M-file T3_2_3, and the results are shown below.

```
>> T3_2_3
The augmented matrix
     49/36      1/2      -1/3      1        1      0
     29/36      1/3      -1/2      2/3      1      0
     46/9       2        -1/3      1        1      0

The row reduced matrix
     1          0         0        30/7     18/7    0
     0          1         0        -75/7    -45/7   0
     0          0         1        -11/7    -15/7   0
```

Interpret the row reduced matrix to arrive at a parametric form of the solution.

$$A = -\frac{30}{7}s - \frac{18}{7}t,\ B = \frac{75}{7}s + \frac{45}{7}t,\ C = \frac{11}{7}s + \frac{15}{7}t,\ D = s,\ E = t$$

Any multiple of 7 for both s and t removes fractions and leads to an integer solution. One choice is $s = -7$ and $t = 7$, and an integer solution is $A = 12$, $B = -30$, $C = 4$, $D = -7$, and $E = 7$. It is useful to check our answer to see if it really is a solution. This is done by forming the dot product of a solution vector with each row of the augmented matrix, excluding the last column. Examine the M-file for details on two checks for the correct solution. The remaining output is

```
A solution x determined from an examination of the
row reduced form is given in row form as
     12    -30     4     -7     7

Floating point format solution check
The dot product of x with each row of M should be zero.
They may not be zero because of roundoff errors.
   2.6645e-015
   1.7764e-015
  -4.4409e-015


*** *** *** *** *** *** *** ***
A solution check when the augmented matrix
is first converted to a symbolic object.
```

The same dot products should be zero.
0
0
0

Numbers in the floating point check are not exactly zero. This is due to roundoff errors in the arithmetic operations. Small numbers of this size are not uncommon after such calculations, and the context of the problem suggests we interpret them as being zero for confirmatory work.

T4. In an experiment for the design of an aircraft wing, the lifting force on the wing is measured at various forward velocities to be:

Velocity (100 ft/s)	1	2	4	8	16	32
Lifting Force (100lb)	0	3.12	15.86	33.7	81.5	123.0

Find an interpolating polynomial of degree 5 that models the data, and use your polynomial to estimate the lifting force at 2,000 ft/s.

Except for the data, this problem is much like the interpolation problem in T1. A partial listing of the M-file T4_2_3 speaks for itself. The function polyval is part of MATLAB, and it evaluates a polynomial specified by a coefficient vector p at points specified in the second argument. Consult help polyval for more details.

```
% T4_2_3
% velocity/force data. Fit a polynomial to
% given data and evaluate the polynomial at
% a specified point.

v = [1,2,4,8,16,32]; % long way
v = 2.^(0:5); % short way

f = [0, 3.12, 15.86, 33.7, 81.5, 123];

n = length(v); % number of points
% The third argument in polyfit indicates
% the degree of the polynomial fit. This
% argument is essential for interpolation.
```

43

```
% Check doc polyfit or help polyfit for details.
p = polyfit(v,f,n-1); % determine a polynomial specified by v and f
%evaluate p at 2000.
val = polyval(p,2000);
```

The output describes some of the intermediate information associated with this problem.

```
>> T4_2_3
The polynomial at v = 2000 is
 -1.6874e+013

The velocity/force values; velocity in first row
    1.0000     2.0000     4.0000     8.0000    16.0000    32.0000
         0     3.1200    15.8600    33.7000    81.5000   123.0000

Polynomial coefficients - highest power down to the
constant term as you read from left to right.
   -0.0005     0.0323    -0.6007     4.2414    -5.8672     2.1948
```

T5. Polynomials other than Taylor polynomials can be used to approximate functions.

    (a) Devise a method for approximating $\sin x$ on the interval $0 \le x \le \pi/2$ by a cubic polynomial.

    (b) Compare the value of $\sin(0.5)$ to the approximation produced by your polynomial.

    (c) Generate the graphs of $\sin x$ and your polynomial over the interval $0 \le x \le \pi/2$ to see how they compare.

To make matters simple, the sine function is sampled at four points $0, \dfrac{\pi}{6}, \dfrac{\pi}{4}, \dfrac{\pi}{2}$ where the values of $\sin(x)$ are known from geometry. The value $\sin(0.5)$ mentioned in part (b) is calculated using an approximation algorithm in MATLAB. It is accurate to around 15 or 16 decimal places. A simultaneous plot of $\sin x$ and the interpolating polynomial $p(x)$ is not recommended because the resolution of the computer screen will cause both to look almost the same in this case. A better way to compare the two is to examine the difference $p(x) - \sin(x)$, and this is what is shown when the M-file T5_2_3 is executed. The M-file is listed here so that you can see how the file is commented and the graphs created and displayed.

```
% T5_2_3
% Polynomial approximation to sin(x) on the interval
% [0, pi/2]; Pick four equally spaced values
% Select values of 0, pi/6, pi/4, and pi/2 for the
% interpolating polynomial. From trigonometry,
% sin([0,pi/6, pi/4,pi/2]) yields [0,.5, 1/2^.5,1]

xvals = [0; pi/6; pi/4; pi/2]; %column vector
sine = [0; .5; 1/2^.5; 1]; % sin(xvals)
% p(x) = a0 + a1*x + a2*x^2 + a3*x^3
% unknowns are [a0; a1; a2; a3]
% Set up 4 linear equations by insisting that
% p(xval(k)) = sin(k), k = 1, 2, 3, 4
% Build the augmented matrix A by columns.
% Use column concatenation
A = [ones(4,1), xvals, xvals.^2, xvals.^3, sine];
R = rref(A);
disp('The row reduced form of the augmented matrix is')
disp(R); % R is a 4 by 5 matrix
% The unknowns are in the last column.

%Use polyval. This requires that the coefficients be listed
%from highest order to lowest order.

p = [R(4,5); R(3,5);R(2,5); R(1,5)];
%p = R(4:-1:1,5); % Here is a cleaner way.
pval = polyval(p,0.5); % polynomial approximation at x = 0.5;
sval = sin(0.5); % A better approximation good to 16 decimals.
% display pval along with sin(0.5);
format short
disp('    p(0.5)    sin(0.5)');
disp([pval, sval]);
disp('The relative percentage error is')
disp(100*abs((pval-sval)/sval));

%Graph both. Use a dashed line for the polynomial and
%a solid line for the sin curve. Add a title.

t = linspace(0,pi/2);% 100 values between 0 and pi/2;
pt = polyval(p,t); % polynomial values
st = sin(t); % sin function at t-values.
plot(t,pt,'k:',t,st,'k');
title('Polynomial Approximation For sin On [0,pi/2]')
xlabel('x');
ylabel('y');
% The curves are so close that they can't be distinguished.
Pause(4) % pause 4 seconds before continuing
% Plot the difference so that you can see the error.
% Note the scaling factor of 10^-3 on the vertical axis.
plot(t,pt-st,'k',xvals,0,'ko'); % stands for black, o is the plot mark
title('Error Plot of p(x) - sin(x) over [0,pi/2]')
xlabel('x'); % add labels to the graph
ylabel('y');
```

45

The output is not long and the graphs are shown below.

```
>> T5_2_3
The row reduced form of the augmented matrix is
    1.0000         0         0         0         0
         0    1.0000         0         0    1.0142
         0         0    1.0000         0   -0.0497
         0         0         0    1.0000   -0.1214

    p(0.5)    sin(0.5)
    0.4795    0.4794

The relative percentage error is
    0.0192
```

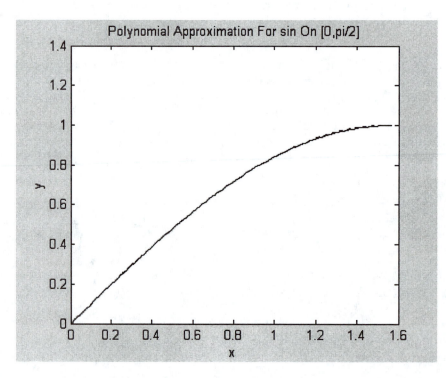

It is much easier to see a difference when p(x) – sin(x) is plotted in the next figure. The error grows as x increases, even though the interpolation rule gives zero error at the four interpolation points. Take note of the scaling factor 10^{-3} in the upper left corner of the error plot below.

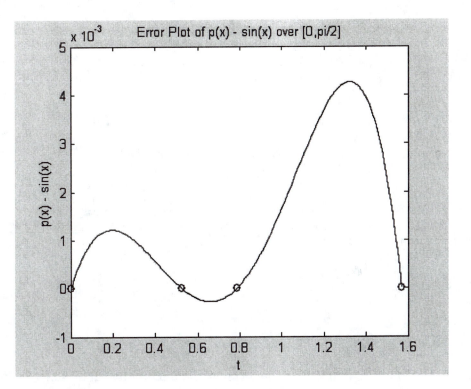

T6. Obtain Formula (20) of Example 8.

Verify a polynomial given in (20) in example 8, page 74, dealing with interpolation. Approximate the integrand with an interpolating polynomial and use five equally spaced values from the interval [0,1] and include the endpoints.

$$\int_0^1 \sin\left(\frac{\pi x^2}{2}\right) dx$$

The command window output below offers verification of the interpolating polynomial described in (20).

```
>> T6_2_3
     Function and Interpolating Polynomial Values
        x                sin(0.5*pi*x^2)           p(x)
                    0                      0   0.00000000000000
    0.25000000000000      0.09801714032956   0.09801714032956
    0.50000000000000      0.38268343236509   0.38268343236509
    0.75000000000000      0.77301045336274   0.77301045336274
    1.00000000000000      1.00000000000000   1.00000000000000
```

47

```
Polynomial coefficients from highest order to lowest

 -2.00543765950561
  2.14427997999068
  0.76236220468845
  0.09879547482649
  0.00000000000000
```

The polynomial representation appears different when polyrep is used.
```
***    ***    ***    ***
p(x) =   -2.0054x^4 + 2.1443x^3 + 0.76236x^2 + 0.098795x + 9.5952e-017
***    ***    ***    ***
```
This happens because the computed constant term from polyfit
is not zero but instead is so small that it appears as zero
when the floating point number is formatted. What you see is
not always what you have.

There is no apparent explanation for the slight discrepancy in the output of polyfit. It is small but puzzling.

T7. Use the method of Example 8 to approximate the integral by subdividing the interval into five equal parts and using an interpolation polynomial to approximate the integrand. Compare your answer to that produced using the numerical integration capability of your calculating utility. The integral is $\int_0^1 e^{x^2} dx$

The M-file for this problem is similar to T6_2_3, and the MATLAB function quadl (voiced as "quad-L") is used to obtain an accurate numerical approximation for a definite integral. The integral in question is approximated using one command. Limits of integration and a character string expression for the function must be provided. This function quadl is discussed again in Chapter 9 when reasons for the string 'exp(x.^2)' are given.

```
>>Q = quadl('exp(x.^2)',0,1); % 0 and 1 are limits of integration.
>>disp(Q)
   1.4627
```

The results show that the approximation is accurate to about 4 decimals. A relative percentage error is reported at the end.

```
>> T7_2_3
         x                      f(x)                    p(x)
                  0    1.00000000000000    1.00000000000000
   0.25000000000000    1.06449445891786    1.06449445891786
   0.50000000000000    1.28402541668774    1.28402541668774
   0.75000000000000    1.75505465696030    1.75505465696030
   1.00000000000000    2.71828182845905    2.71828182845905

Polynomial coefficients from highest order to lowest
    1.53853722744385
   -1.27888014889215
    1.52634206547861
   -0.06771731557126
    1.00000000000000

Using polyrep, p(x) = 1.5385x^4 - 1.2789x^3 + 1.5263x^2 - 0.067717x + 1

Integral of Interpolating Polynomial
    1.46290943897297

MATLAB built-in Result
    1.46265176347799

Percentage Relative Error
    0.01761700914835
```

The next problem is fairly involved. It offers an example of how organization before computing leads to a documented M-file that describes the numerical algorithms needed to solve a long problem. It is a little awkward because rref is the main tool used for computations.

T8. Suppose that a ship with unknown coordinates (x, y, z) at an unknown time t receives the following data from four satellites.

Satellite	Satellite Position	Time
1	(0.94, 1.2, 2.3)	1.23
2	(0, 1.35, 2.41)	1.08
3	(2.3, 0.94, 1.2)	0.74
4	(2.41, 0, 1.35)	0.25

Using the technique of Example 1, find the coordinates of the ship and verify that it is a unit distance from the earth's center.

Consult the text at this point so that you can follow the rather lengthy calculations. A strategy is to look for patterns by using symbols instead of numbers. Numbers can be substituted for symbols once the patterns are discovered. This is how you discover linear systems to be solved.

Set $w = .47^2 = 0.2208$. Associate a point $p_1 = (x_1, y_1, z_1) = (0.94, 1.2, 2.3)$ with the given data and start with the condition determined by the first point

$$(x - x_1)^2 + (y - y_1)^2 + (z - z_1)^2 = w(t - t_1)^2$$

Expand the quadratic terms

$$x^2 - 2x_1 x + x_1^2 + y^2 - 2y_1 y + y_1^2 + z^2 - 2z_1 z + z_1^2 = w\left(t^2 - 2t_1 t + t_1^2\right)$$

Introduce simplifying expressions $u = x^2 + y^2 + z^2 - wt^2$, $u_1 = x_1^2 + y_1^2 + z_1^2 - wt_1^2$ and simplify further

$$2x_1 x + 2y_1 y + 2z_1 z - 2wt_1 t = u + u_1$$

This calculation is carried out for the three remaining points using $p_2 = (x_2, y_2, z_2) = (0, 1.35, 2.41)$, and so on. The relationships

$$(x - x_k)^2 + (y - y_k)^2 + (z - z_k)^2 = w(t - t_k)^2, \quad k = 1, 2, 3, 4$$

lead to

$$2x_k x + 2y_k y + 2z_k z - 2wt_k t = u + u_k, \quad k = 1, 2, 3, 4$$

where

$$u_k = x_k^2 + y_k^2 + z_k^2 - wt_k^2$$

This set of equations is a symbolic version of four similar equations in the problem description. As described in the text, successively subtract the second third, and fourth equations from the first one to obtain a linear system of three equations in four unknowns. Write the equations as

$$(x_1 - x_k)x + (y_1 - y_k)y + (z_1 - z_k)z - w(t_1 - t_k)t = \frac{u_1 - u_k}{2}, k = 2, 3, 4$$

and introduce the symbolic 3x4 augmented matrix for this system

$$\begin{bmatrix} (x_1 - x_2) & (y_1 - y_2) & (z_1 - z_2) & -w(t_1 - t_2) & 0.5(u_1 - u_2) \\ (x_1 - x_3) & (y_1 - y_3) & (z_1 - z_3) & -w(t_1 - t_3) & 0.5(u_1 - u_3) \\ (x_1 - x_4) & (y_1 - y_4) & (z_1 - z_4) & -w(t_1 - t_4) & 0.5(u_1 - u_4) \end{bmatrix}$$

It is time to use the data to generate a data matrix that can be row reduced. There are three equations and four unknowns, and we expect to get a parametric form of the solution, just as in Example 1. Patterns listed in the augmented matrix suggest that we organize the point and time data as column vectors. Once numerical values replace the symbols, the augmented matrix can be row reduced using `rref`. Imagine that is done. For purposes of discussion, let R denote the row reduced form of the matrix and use the notation r_{ij} for the i, j element in R. Looking ahead, the row reduced form suggests that we express x, y, z, in terms of t. The symbolic solution is

(**) $x = -r_{14}t + r_{15}$, $y = -r_{24}t + r_{25}$, and $z = -r_{34}t + r_{35}$

Writing the solution this way lets us use the results of the row reduction process without having to copy numbers down before proceeding. Substitute into the first equation

$$(-r_{14}t + r_{15} - x_1)^2 + (-r_{24}t + r_{25} - y_1)^2 + (-r_{34}t + r_{35} - z_1)^2 = w(t - t_1)^2$$

Anticipating subsequent calculations, introduce $a = x_1 - r_{15}$, $b = y_1 - r_{25}$, and $c = z_1 - r_{35}$. Simplify as

$$(r_{14}t + a)^2 + (r_{24}t + b)^2 + (r_{34}t + c)^2 = w(t - t_1)^2$$

Expand each quadratic by hand and rearrange the terms for a quadratic polynomial

$$(r_{14}^2 + r_{24}^2 + r_{34}^2 - w)t^2 + 2(ar_{14} + br_{24} + cr_{34} + wt_1)t + (a^2 + b^2 + c^2 - wt_1^2) = 0$$

51

The numerical coefficients for the polynomial are obtained from this expression, the data, and the row reduced matrix R for the system. Once the roots of this polynomial are known, it is possible to calculate the location of the ship using (**) and verify that it is about 1 earth radius from the center. This completes an outline of the solution process.

The M-file T8_2_3 that carries out this complex series of calculations is too long to list here. It includes extensive comments so that you can review the organization. Section 1 contains the data for the problem, while section 2 completes the computation of each u_k. The augmented matrix for the linear system is generated in section 3 using the symbolic matrix template above, and its row reduced form is calculated and displayed in section 4. The three coefficients for the quadratic in t are computed in section 5, and a short snippet of the code shows how they are calculated.

```
a = p1(1) - R(1,5); % x1 - r15
b = p1(2) - R(2,5); % y1 - r25
c = p1(3) - R(3,5); % z1 - r35

q2 = sum(R(1:3,4).^2) - 0.22; % coefficient of t^2 from column 4 of R
q1 = 2*(a*R(1,4) + b*R(2,4) + c*R(3,4) + 0.22*t(1)); % coefficient of t
q0 = a^2 + b^2 + c^2 - 0.22*t(1,1)^2;% constant term
q = [q2, q1, q0]; % quadratic polynomial, MATLAB style
```

The location of the ship is calculated in section 6 along with the distance from the earth's center. Results of these calculations are displayed next using a short format. The calculations show that t is about 5.1569 and the distance from the center is about 0.9926 earth radii. The problem uses 0.22 instead of 0.2201 for one of the parameters. It is interesting to see how the results change when 0.22 is used. The M-file has a note in section 2 where you can change one line of code and see the new results. It shows how sensitive the solution is to small changes in the data. Be sure you remember to comment the line again when you are finished so that you retain the best data.

```
Augmented matrix A
    0.9400   -0.1500   -0.1100   -0.0331   -0.0468
   -1.3600    0.2600    1.1000   -0.1082   -0.1066
   -1.4700    1.2000    0.9500   -0.2209   -0.1698

Row reduced matrix R
```

```
   1.0000          0          0    -0.0785    -0.0878
        0     1.0000          0    -0.1544    -0.1062
        0          0     1.0000    -0.1589    -0.1804
```

Roots of quadratic for t
  -10.0458
    5.1569

Ship Location
        x   y   z
    0.3169   0.6900   0.6393

Distance from calculated position to the center of the earth
and expressed in earth radii.
    0.9926

It can be argued that the effort put into this problem is hardly worth the results. There are several positive aspects of this approach. The documentation in the M-file provides a fairly clear picture of the process of converting symbolic results into algorithms that are used to calculate numbers. Also, the M-file is organized so that other data could also be used instead of the suggested numbers. You decide.

Command Summary

solvesys(A,b) --- report generator for solving $Ax = b$ for rectangular A

rref   --- rref(A) row reduces A

rrefsym   --- [R,cols] = rrefsym(A) returns pivot columns too.

polyfit   --- find the coefficients of an interpolating polynomial

polyrep   --- convert a vector into a string representation for a polynomial

conv   --- conv(p,q) is another way to multiply polynomials p and q

quadl   --- integrate a function specified by a character vector

plot   --- plot data. This is a powerful command

xlabel   --- put a label on the horizontal axis of a plot

ylabel   --- put a label on the vertical axis of a plot

solve   --- symbolic routine for solving equations

subs   --- substitution command for symbolic expressions

sum   --- sum(x) adds the elements of x

Chapter 3

Matrices and Matrix Algebra

EXERCISE SET 3.1
TECHNOLOGY EXERCISES

T1. (*Matrix addition and multiplication by scalars*)   Perform the calculations in Examples 2 and 3.

(a) Matrices are taken from Example 2, page 80.

Three different matrix entry styles are illustrated in T1a_3_1.

```
% T1a_3_1
%Verify certain matrix operations from example 2 in section 3.1
A = [2,1,0,3;-1,0,2,4;4,-2,7,0];% Use commas and semi-colons. Safe.
% The next style omits the semi-colon but uses a carriage return to
% separate rows. MATLAB recognizes this style of input.
B = [-4, 3, 5, 1
     2, 2, 0, -1
     3, 2, -4, 5];
% The last style uses no commas, but make sure there is a space
% between the numbers. Not recommended for complicated
% data entry.
C = [1 1;2 2];
%
disp('A is'); disp(A)
disp('B is'); disp(B)
disp('C is'); disp(C)
disp('*****')
disp('A+B is'); disp(A+B)
disp('A-B is'); disp(A-B)
disp('A+C is'); disp(A+C) % This generates an error message
disp(B+C) % These give error messages
disp(A-C) % similar to disp(A+C)
disp(B-C)
```

A partial listing output of the command window output provides basic details.

```
>> T1a_3_1
A is
     2     1     0     3
    -1     0     2     4
     4    -2     7     0

C is
     1     1
     2     2
```

```
*****
A-B is
     6    -2    -5     2
    -3    -2     2     5
     1    -4    11    -5

A+C is
??? Error using ==> +
Matrix dimensions must agree.

Error in ==> C:\WINDOWS\DESKTOP\codes\T1a_3_1.m
On line 20   ==> disp('A+C'); disp(A+C)
```

Execution of an M-file ceases when an error is encountered. A brief message similar to the one above gives some idea where the problem occurred. Errors later in the M-file are not detected until you make a correction and start over. This is one reason why M-files are so useful – only a text file has to be edited and saved.

(b) Use matrices from Example 3 and carry out the same operations as in part (a).

Similar calculations are carried out in T1b_3_1 using the matrices in Example. No output is shown.

T2. (***Matrix multiplication***) Compute the product in Example 5.

Executing this M-file displays the results of the 2 by 4 product matrix *AB*.

```
% T2_3_1
%Verify a matrix product from example 5 in section 3.1
A = [1 2 4;2 6 0]; % data entry by rows
B = [4 1 4 3;0 -1 3 1;2 7 5 2];

disp('A is');disp(A);
disp('B is');disp(B);
disp('*****')
disp('AB is'); disp(A*B)
```

T3. (***Trace and Transpose***) Find the trace and the transpose of the matrix

$$A = \begin{bmatrix} 3 & 1 & -7 \\ 2 & 4 & 11 \\ 3 & 3 & 9 \end{bmatrix}$$

A listing of the M-file followed by the resulting output gives some idea how the echo command functions.

```
% T3_3_1
%Calculate a transpose and a trace of A
A = [3 1 -7;2 4 11;3 3 9];

% echo on causes subsequent lines and the resulting output appear
% in the command window.
echo on
disp(A)
% *****
% A' evaluates the transpose of A
disp(A');
% transpose(A) is another way to get the transpose
disp(transpose(A))
% trace of A, tr(A)
disp(trace(A))
%MATLAB syntax is trace(A), writing syntax is tr(A)
echo off % turn off the echo feature
```

The command line output shows the results of the calculations.

```
>> T3_3_1
disp(A)
      3      1     -7
      2      4     11
      3      3      9
% *****
% A' evaluates the transpose of A
disp(A');
      3      2      3
      1      4      3
     -7     11      9
% transpose(A) is another way to get the transpose
disp(transpose(A))
      3      2      3
      1      4      3
     -7     11      9
% trace of A, tr(A)
disp(trace(A))
     16
%MATLAB syntax is trace(A), writing syntax is tr(A)
echo off % turn off the echo feature
```

T4. (*Extracting row vectors, column vectors, and entries*)

(a) Extract the row vectors and column vectors of the matrix in Exercise T3

(b) Find the sum of the row vectors and the sum of the column vectors of the matrix in Exercise T3.

(c) Extract the diagonal entries of the matrix in Exercise T3 and compute their sum.

The M-file listing shows how the colon operator ':' makes it easy to extract rows and columns from a matrix.

```
% T4_3_1
% Extract rows, columns, and diagonal
% from the matrix of exercise T3.
A = [3 1 -7;2 4 11;3 3 9];

disp('A'); disp(A)
disp('*** extract rows ***')
disp('row 1');disp(A(1,:));
disp('row 2');disp(A(2,:));
disp('*** extract columns ***')
c1 = A(:,1); % extracted columns or rows can be assigned.
disp('column 1'); disp(c1);
disp('column 2'); disp(A(:,2));
disp('*** add row vectors ***')
disp(A(1,:) + A(2,:) + A(3,:))
%disp(sum(A)) % another way
disp('*** add column vectors ***')
disp(A(:,1) + A(:,2) + A(:,3));
%disp(sum(A,2)); %another way.
% 2 signals to add across the second coordinate
% and this means add the columns.

disp('*** Extract the diagonal of A ***')
disp('Use the diag operator')
d = diag(A); % extracted as a column vector
disp(d)
```

Command window output is not listed here, and the vectors created in the M-file can be examined to reinforce the presentation.

T5. See what happens when you try to multiply matrices whose sizes do not conform for the product.

The simplest of matrices illustrates the effect.

```
>> C = [1 2]; % row vector
>> D = C*C;
??? Error using ==> *
Inner matrix dimensions must agree.
```

Calculations such as these cause all execution to stop.

T6. (*Linear combinations by matrix multiplication*). One way to obtain a linear combination $c_1\mathbf{v}_1 + c_2\mathbf{v}_2 + \cdots + c_k\mathbf{v}_k$ of vectors in R^n is to compute the matrix $A\mathbf{c}$ in which the successive column vectors of A are $\mathbf{v}_1, \mathbf{v}_2, \ldots, \mathbf{v}_k$ and $\mathbf{c}$ is the column vector whose successive entries are $c_1, c_2, \ldots, c_k$. Use this method to compute the linear combination

$$6(8, 2, 1, 4) + 17(3, 9, 11, 6) + 9(0, 1, 2, 4)$$

The organization of the data and the resulting linear combination are explained in the M-file.

```
% T6_3_1
% Form linear combinations using a matrix/vector product

c = [6 17 9]'; % column vector because of transpose.
v1 = [8 2 1 4]'; % column vector
v2 = [3 9 11 6]';
v3 = [0 2 2 4]';
A = [v1, v2, v3];% concatenate to build A
oneway = c(1)*v1 + c(2)*v2 + c(3)*v3;% linear combination
anotherway = A*c;
disp('Vectors as columns of A')
disp(A)
disp('c'' is displayed instead of c')
disp(c')
disp('c(1)*v1 + c(2)*v2 + c(3)*v3 is')
disp(oneway);
disp('Linear combination computed as A*c')
disp(anotherway)
```

Subsequent output verifies the request.

```
>> T6_3_1
Vectors as columns of A
     8     3     0
     2     9     2
     1    11     2
     4     6     4

c' is displayed instead of c
     6    17     9

c(1)*v1 + c(2)*v2 + c(3)*v3 is
    99
   183
   211
   162

Linear combination computed as A*c
```

```
    99
   183
   211
   162
```

T7 Use the idea in Exercise T6 to compute the following linear combinations with a single matrix multiplication.

$$3(7, 1, 0, 3) - 4(-1, 5, 7, 0) + 2(6, 3, -2, 1)$$

$$5(7, 1, 0, 3) - (-1, 5, 7, 0) + (6, 3, -2, 1)$$

$$2(7, 1, 0, 3) + 4(-1, 5, 7, 0) + 7(6, 3, -2, 1)$$

The strategy is to build column vectors from each vector listed and then create A by concatenating the columns. Weights(multipliers) are packaged as a column vector called c. The linear combination is realized as Ac, and a partial listing of the M-file T7_3_1 shows how the results are calculated.

```
% T7_3_1
% Form linear combinations using a matrix/vector product
% Build the matrix from T6
v1 = [8 2 1 4]'; % column vector
v2 = [3 9 11 6]';
v3 = [0 2 2 4]';
A = [v1, v2, v3];% concatenate to build A
c = [3 -4 2]'; % column vector of weights
LC = A*c; % linear combination)
% The rest of the M-file is not listed here.
```

Most of the M-file is devoted to formatting results so that they can be read in the command window, and a partial listing of the output is given below.

```
>> T7_3_1
Vectors as columns of A
      8      3      0
      2      9      2
      1     11      2
      4      6      4

*** first case ***
c' is displayed instead of c
      3     -4      2

Linear combination in column form: A*c
     12
    -26
```

```
    -37
    -4
```

EXERCISE SET 3.2
TECHNOLOGY EXERCISES

T1. (*Matrix powers*) Compute various positive powers of the matrix

$$A = \begin{bmatrix} 1 & 2 & -3 & 0 \\ 1 & 1 & -2 & 1 \\ 2 & 1 & 3 & 4 \\ -3 & 2 & 2 & -8 \end{bmatrix}$$

The syntax A^k evaluates the kth power of a square matrix.

```
>> disp(A)
     1      2     -3      0
     1      1     -2      1
     2      1      3      4
    -3      2      2     -8
>> disp(A^2)  % A^2 = A*A
    -3      1    -16    -10
    -5      3     -9    -15
    -3     16      9    -19
    27    -18     -5     74
```

Additional calculations are given in T1_3_2.

T2. Compute $A^5 - 3A^3 + 7A - 4I$ for the matrix A in Exercise T1.

Including the identity matrix is the most important part of this problem and eye(n) generates an identity matrix of size n. Details are given in T2_3_2. Evaluating the polynomial can be done by hand or with a MATLAB command polyvalm. Do not confuse this command with polyval. The code fragments below indicate what happens.

```
>> B = diag([2 3]);
>> disp(B)
     2      0
     0      3
>> disp(polyval([1 1],B)) % it works
     3      1
     1      4
>> disp(polyvalm([1 1],B)) % 1*B + 1*eye(2)
```

```
    3      0
    0      4
>> disp(B + 1)  % this shows how polyval treats the constant term
    3      1
    1      4
```

MATLAB treats a matrix plus a number in a useful way that is initially confusing. In effect,

$$B + 1 \leftrightarrow \begin{bmatrix} 3 & 0 \\ 0 & 4 \end{bmatrix} + \begin{bmatrix} 1 & 1 \\ 1 & 1 \end{bmatrix}$$

This is why the command `polyval(v,A)` subtracted `4*ones(n)` rather that `4*eye(n)`.

T3. Confirm Formulas (12) and (13) for

$$A = \begin{bmatrix} 1 & 2 & -3 & 0 \\ 1 & 1 & -2 & 1 \\ 2 & 1 & 3 & 4 \\ -3 & 2 & 2 & -8 \end{bmatrix}, \quad \mathbf{u} = \begin{bmatrix} -1 \\ 2 \\ 3 \\ 5 \end{bmatrix}, \quad \mathbf{v} = \begin{bmatrix} 2 \\ 0 \\ -4 \\ 1 \end{bmatrix}$$

These formulas refer to matrix products and transposes, and the M-file T3_3_2 generates and displays the results. No listings are given.

T4. Let

$$A = \begin{bmatrix} 0 & \frac{1}{2} & \frac{1}{3} \\ \frac{1}{4} & 0 & \frac{1}{5} \\ \frac{1}{6} & \frac{1}{7} & 0 \end{bmatrix}$$

Discuss the behavior of A^k as k increases indefinitely; that is, as $k \to \infty$.

Powers of a matrix are simple to compute in MATLAB, and only a partial listing of the output of T4_3_2 is shown. Note the scaling factor 1.0e-014 $= 10^{-14}$ that multiplies each element of A^50 in the output. This is used for display purposes and it seems clear that every entry of A^k approaches zero as k gets large.

```
>> T4_3_2
A
        0          1/2        1/3
        1/4        0          1/5
        1/6        1/7        0

*** Powers of A **
A^50
   1.0e-014 *

        0.1491     0.1925     0.1714
        0.1023     0.1320     0.1176
        0.0767     0.0990     0.0882
```

T5. An idempotent matrix is one that satisfies $A^2 = A$.

(a) Show that the following matrix is idempotent (see Exercise 32):

$$A = \begin{bmatrix} \frac{1}{3} & -\frac{2}{3} & -\frac{2}{3} \\ -\frac{2}{3} & \frac{1}{3} & -\frac{2}{3} \\ -\frac{2}{3} & -\frac{2}{3} & \frac{1}{3} \end{bmatrix}$$

(b) Confirm the statements in parts (a) and (b) of Exercise32.

That A is idempotent is verified in the M-file T5_3_2. Properties of $I - A$ and $2A - I$ are also confirmed. No new ideas are needed for this problem.

T6. A square matrix A is said to be *nilpotent* if $A^k = 0$ for some positive integer k. The smallest value of k for which this equation holds is called the *index of nilpotency*. In each part, confirm that the matrix is nilpotent, and find the index of nilpotency.

(a) $A = \begin{bmatrix} 0 & 0 & 0 & 0 \\ 1 & 0 & 0 & 0 \\ 2 & 1 & 0 & 0 \\ -3 & 2 & 2 & 0 \end{bmatrix}$ (b) $B = \begin{bmatrix} 0 & 0 & 1 & 0 \\ 1 & 0 & 0 & 0 \\ 0 & 1 & 0 & 0 \\ 0 & 0 & 0 & 1 \end{bmatrix}$

The calculations in T6_3_2 show that A is nilpotent because $A^3 \neq 0$ and $A^4 = 0$. The same file shows that B is nilpotent because $B^3 \neq 0$ and $B^4 = 0$. No output is shown.

T7. (CAS) Make a conjecture about the form of A^n for positive integer powers of n.

$$\text{(a)} \quad A = \begin{bmatrix} a & 1 \\ 0 & a \end{bmatrix} \qquad \text{(b)}\, A = \begin{bmatrix} \cos\theta & \sin\theta \\ -\sin\theta & \cos\theta \end{bmatrix}$$

A short experiment gives the following results.

```
>> T7_3_2
A
[ a, 1]
[ 0, a]

A^3
[    a^3, 3*a^2]
[      0,   a^3]

A^5
[    a^5, 5*a^4]
[      0,   a^5]
```

It appears that $A^k = a^k \begin{bmatrix} 1 & 0 \\ 0 & 1 \end{bmatrix} + ka^{k-1} \begin{bmatrix} 0 & 1 \\ 0 & 0 \end{bmatrix}$, and code is given in T7_3_2.

Part two of the problem asks for simplified versions of A^k when

$$A = \begin{bmatrix} \cos\theta & \sin\theta \\ -\sin\theta & \cos\theta \end{bmatrix}.$$

The symbolic Toolbox command $\mathtt{simple}$ takes care of the trigonometric simplifications for this problem, and only part of the output of T7_3_2 is shown here.

```
>> T7_3_2
some output is not shown
B
[  cos(t),  sin(t)]
[ -sin(t),  cos(t)]
```

```
B^2 is not simplified yet.
[ cos(t)^2-sin(t)^2,    2*cos(t)*sin(t)]
[  -2*cos(t)*sin(t), cos(t)^2-sin(t)^2]

A simplified version of B^2 looks better.
[  cos(2*t),  sin(2*t)]
[ -sin(2*t),  cos(2*t)]

A simplified version of B^3 using the command simple is
[  cos(3*t),  sin(3*t)]
[ -sin(3*t),  cos(3*t)]
```

A general pattern is now apparent, and an induction proof shows that

$$B^m = \begin{bmatrix} \cos(mt) & \sin(mt) \\ -\sin(mt) & \cos(mt) \end{bmatrix}$$

EXERCISE SET 3.3
TECHNOLOGY EXERCISES

T1. (*Inverses*) Compute the inverse of the matrix in Example 3, and see what happens when you try to compute the inverse of the singular matrix in Example 4.

MATLAB reports a failed attempt as found in T1_3_3.

```
>> T1_3_3
 *** Matrix From Example 3 ***
A is
      1      2      3
      2      5      3
      1      0      8
The inverse of A is
  -40.0000    16.0000     9.0000
   13.0000    -5.0000    -3.0000
    5.0000    -2.0000    -1.0000
 *** Singular Matrix B From Example 3 ***
B is
      1      6      4
      2      4     -1
     -1      2      5
Attempt inv(B)
Warning: Matrix is singular to working precision.
> In C:\WINDOWS\DESKTOP\codes\T1_3_3.m at line 19
The result of the attempt is Binv
    Inf    Inf    Inf
    Inf    Inf    Inf
    Inf    Inf    Inf
```

MATLAB indicates where the problem occurred in the M-file. The failed attempt Binv consists of Inf, which stands for infinity as a way of indicating the result. It serves no purpose at this time.

T2. (**Augmented Matrices**) Many Technology utilities provide methods for building up new matrices from a set of specified matrices. Determine whether your utility provides for this, and if so form the augmented matrix for the system in Example 5 from the matrices **A** and **b**.

Matrices have been augmented since Chapter 1, and there is nothing new.

```
>> T2_3_3
A is
     1     2     3
     2     5     3
     1     0     8
b is
     5
     3
    17
The augmented matrix [A,b] is
     1     2     3     5
     2     5     3     3
     1     0     8    17
```

T3. See what happens when you try to compute a negative power of the singular matrix in Example 4.

Negative powers A^{-m} are equivalent to $\left(A^{-1}\right)^{m}$, and the results are what you would expect since A does not have an inverse. Not output or code is given here. Check the M-file T3_3_3.

T4. Compute the inverse of the matrix A in Example 3 by adjoining the 3×3 identity matrix to A, and reducing the 3×6 matrix to reduced row echelon form.

The command R = rref([A,eye(3)]) augments the identity I to A on the right and then returns the row reduced form in R. The inverse of A is found in the last three columns of R, and the M-file T4_3_3 provides details. The matrix eye(3) is an identity matrix of size 3. No file listing or output is given here.

T5. Solve the following matrix equation for X:

$$\begin{bmatrix} 1 & -1 & 1 \\ 2 & 3 & 0 \\ 0 & 2 & -1 \end{bmatrix} X = \begin{bmatrix} 2 & -1 & 5 & 7 & 8 \\ 4 & 0 & -3 & 0 & 1 \\ 3 & 5 & -7 & 2 & 1 \end{bmatrix}$$

Let A denote the matrix on the left and let B denote the matrix on the right and think of the system as $AX = B$. After the augmented matrix [A,B] is row reduced to R, the solution X is found in the last five columns of R. Another approach is to use the inverse of A to arrive at $X = A^{-1}B$. Details are in the M-file T5_3_3.

T6. (CAS) Obtain Formula (5) of Theorem 3.2.7.

Once symbolic variables are created, the matrix in Theorem 3.2.7 is inverted using the command inv(A). Details are in the M-file T6_3_3, and the listing below verifies the result given in the theorem.

```
>> T6_3_3
A is
[ a, b]
[ c, d]
A^-1 is
[  d/(a*d-b*c),  -b/(a*d-b*c)]
[ -c/(a*d-b*c),   a/(a*d-b*c)]
```

T7. Various methods are used to solve a linear system.

   (a) Use matrix inversion to solve the linear system

$$
\begin{bmatrix} 3 & 3 & -4 & -3 \\ 0 & 6 & 1 & 1 \\ 5 & 4 & 2 & 1 \\ 2 & 3 & 3 & 2 \end{bmatrix} \begin{bmatrix} x_1 \\ x_2 \\ x_3 \\ x_4 \end{bmatrix} = \begin{bmatrix} -2 \\ 3 \\ 5 \\ 1 \end{bmatrix}
$$

(b) Solve the system in part (a) by reducing the augmented matrix to reduced row echelon form.

(c) Solve the system in part (a) using the system solving capability of your utility and compare the result to that obtained parts (a) and (b).

Assign the matrix on the left to A and the column on the right to b. The first two parts are familiar, but it is important to address the third part using the backslash operator in MATLAB. The solution is given by x = A\b and the backslash \ is a MATLAB command to solve $Ax = b$ in a reliable manner using gaussian elimination. It is the preferred way to solve linear systems in MATLAB because the row reduction command rref is not as reliable when used with numeric matrices. The M-file T7_3_3 provides details.

```
>> A = [1 2;3 4];
>> b = [4;6];
>> x = A\b; % use \ to get x
>> disp(x)
     -2
      3
>> disp(A*x - b) % verify that x is a solution
      0
      0
```

T8. By experimenting with different values of n, find an expression for the inverse of an $n \times n$ matrix of the form

$$
A = \begin{bmatrix}
1 & 2 & 3 & 4 & \cdots & n-1 & n \\
0 & 1 & 2 & 3 & \cdots & n-2 & n-1 \\
0 & 0 & 1 & 2 & \cdots & n-3 & n-2 \\
\vdots & \vdots & \vdots & \vdots & & \vdots & \vdots \\
0 & 0 & 0 & 0 & \cdots & 1 & 2 \\
0 & 0 & 0 & 0 & \cdots & 0 & 1
\end{bmatrix}
$$

Look at the patterns in this matrix. In a given row, the first non-zero element is 1 and the subsequent elements are an increasing arithmetic progression. There is a command that produces the related matrix upper triangular matrix C of ones. One example is

$$C = \begin{bmatrix} 1 & 1 & 1 \\ 0 & 1 & 1 \\ 0 & 0 & 1 \end{bmatrix}$$

Matrix A is found by adding up the rows in a cumulative fashion using the cumsum command. A short piece of code illustrates the idea.

```
>> n = 4;
>> C = ones(n,n); % matrix of ones
>> C = triu(C); % retain the upper triangular part
>> A = cumsum(C,2); % form cumulative sum for each row
>> disp(A)
     1     2     3     4
     0     1     2     3
     0     0     1     2
     0     0     0     1
```

The cumsum command forms the cumulative sum of columns or rows of a matrix: cumsum(C) forms cumulative sums for each column and cumsum(C,2) does the same for rows. See help cumsum for details. The M-file T8_3_3 provides experimental results that let you guess the general form of the inverse.

T9. (CAS) The $n \times n$ matrix $H_n = [h_{ij}]$ for which $h_{ij} = 1/(i+j-1)$ is called the nth order *Hilbert matrix*.

(a) Write out the Hilbert matrices H_2, H_3, H_4, and H_5.

(b) Hilbert matrices are invertible, and their inverses can be found exactly using computer algebra systems (if n is not too large). Find the exact inverses of the Hilbert matrices in part (a). [*Note*: Some programs have a command for automatically entering Hilbert matrices by specifying n. If your program has this feature, it will save some typing.]

(c) Hilbert matrices are notoriously difficult to invert numerically because of their sensitivity to slight roundoff errors. To illustrate this, create an approximation H to H_5 by converting the fractions in H_5 to decimals with six decimal places. Invert H and compare the result to the exact inverse you obtained in part (b).

MATLAB provides Hilbert numeric matrices and their inverses. The commands are `hilb(n)` and `invhilb(n)` where n is a positive integer that is not too large. The third part of this problem is not possible with standard MATLAB commands because variable precision is not available unless the Symbolic Toolbox is invoked. There is a way to get at the truncated matrix using a trick: `str2num(num2str(H5,'%9.6f'))`. Look through the M-file for insight.

T10. Let

$$A = \begin{bmatrix} 1 & 3 & 2 \\ 4 & 5 & 1 \\ 3 & 7 & 2 \end{bmatrix}$$

Find a quadratic polynomial $f(x) = ax^2 + bx + c$ for which $f(A) = A^{-1}$.

This idea goes back to Section 3.2 in the text. Multiplication by A gives an equivalent system.

$$f(A) = A^{-1} \leftrightarrow aA^3 + bA^2 + cA = I$$

The matrix on the left is a linear combination of powers of A. Two matrices are equal when all their elements agree. The (1,1) entry of $aA^3 + bA^2 + cA = I$ is $174a + 19b + c = 1$, and the strategy is to repeat this for all the remaining elements and form a linear system of nine equations in 3 unknowns. The resulting overdetermined system has a solution $a = 1/14$, $b = -4/7$, and $c = -4/7$, and the M-file T10_3_3 explains the details.

70

T11. If A is a square matrix, and if $f(x) = p(x)/q(x)$, where $p(x)$ and $q(x)$ are polynomials for which $q(A)$ is invertible, then we define $f(A) = p(A)q(A)^{-1}$. Find $f(A)$ for $f(x) = \dfrac{x^3 + 2}{x^2 + 1}$ and the matrix in Exercise T9.

The strategy is to determine $p(A)$, $q(A)$, and then calculate the matrix product $p(A)q(A)^{-1}$. The output of T11_3_3 gives the result and intermediate calculations.

EXERCISE SET 3.4
TECHNOLOGY EXERCISES

T1. (*Sigma notation*) Use your technology utility to compute the linear combination

$$v = \sum_{j=1}^{25} c_j v_j$$

for $c_j = 1/j$ and $v_j = (\sin j, \cos j)$

This sum is a matrix product Ac in disguise, where $A = \begin{bmatrix} \sin(1) & \sin(2) & \cdots & \sin(35) \\ \cos(1) & \cos(2) & \cdots & \cos(35) \end{bmatrix}$

and $c = (1:35)'$. Review the M-file T1_3_4 for details.

T2. Devise a procedure for using your technology utility to determine whether a set of vectors in R^n is linearly independent, and use it to determine whether the following vectors are linearly independent.

$v_1 = (4, -5, 2, 6)$, $v_2 = (2, -2, 1, 3)$, $v_3 = (6, -3, 3, 9)$, $v_4 = (4, -1, 5, 6)$

One strategy is to form a matrix A whose columns are the given vectors and then check if $Ax = 0$ has non-trivial solutions. From an operational standpoint, the row reduced form of A tells enough to allow a decision. The vectors are linearly independent if each column of the row reduced form is a pivot column. Details, results, and a cautionary note are given in the M-file T2_3_4.

71

T3. Let $v_1 = (4, 3, 2, 1)$, $v_2 = (5, 1, 2, 4)$, $v_3 = (7, 1, 5, 3)$, $x = (16, 5, 9, 8)$, $y = (3, 1, 2, 7)$. Determine whether x and y lie in span$\{v_1, v_2, v_3\}$.

Spanning is equivalent to the question of the consistency of the systems $Au = x$ and $Aw = y$, where the columns of A are the vectors v_k. Both questions can be answered by row reducing $[A, x]$ and $[A, y]$, and then interpreting the results based on the pivot columns in the row reduced forms. This is carried out in the M-file T3_3_4.

EXERCISE SET 3.5
TECHNOLOGY EXERCISES

T1. (a) Show that the vector $v = (-21, -60, -3, 108, 84)$ is in span$\{v_1, v_2, v_3\}$, where

$v_1 = (1, -1, 3, 11, 20)$, $v_2 = (10, 5, 15, 20, 11)$, and $v_3 = (3, 3, 4, 4, 9)$.

(b) Express v as a linear combination of v_1, v_2, and v_3.

It is shown in the M-file T1_3_5 that $v = 12v_1 + 3v_2 - 21v_3$. This is accomplished by finding the weight vector x for the system $\begin{bmatrix} v_1, & v_2, & v_3 \end{bmatrix} x = v$ as found in the row reduced form.

EXERCISE SET 3.6
TECHNOLOGY EXERCISES

T1. (*Special Types of Matrices*) Typing in the entries of a matrix is tedious, so many technology utilities provide shortcuts for entering identity matrices, zero matrices, diagonal matrices, triangular matrices, symmetric matrices, and skew symmetric matrices. Determine whether your utility has this feature, and, if so, practice entering matrices of various special types.

MATLAB has special commands for matrices of zeros, ones, diagonal matrices and triangular matrices. There are no special commands for generating symmetric or skew symmetric matrices. The M-file T1_3_6 describes and illustrates many of them. There are

72

also matrices that are tedious to enter and have interesting properties. The MATLAB names `magic`, `hilb`, `invhilb`, `hadamard`, and `vander` are available. Do a search for `vander` in the help files for more details on this function. See the appropriate help file for details on the others. Magic matrices are interesting. Some output is shown here.

```
I3 = eye(3); % 3x3 identity matrix
disp(I3)
     1     0     0
     0     1     0
     0     0     1
onesquare = ones(3); % 3x3 matrix of ones.
disp(onesquare);
     1     1     1
     1     1     1
     1     1     1

onerect = ones(3,4); % 3x4 matrix of ones.
disp(onerect)
     1     1     1     1
     1     1     1     1
     1     1     1     1

U = triu(onesquare); % extract upper triangular part
disp(U)
     1     1     1
     0     1     1
     0     0     1
```

T2. Confirm the results in Theorem 3.6.1 for some triangular matrices of your choice.

An extensively documented M-file T2_3_6 demonstrates many of the results in Theorem 3.6.1 that refer to properties of triangular matrices.

T3. Show that the matrix A is nilpotent, and then use Formula (12) to compute $(I - A)^{-1}$. Check your answer by computing the inverse directly.

$$A = \begin{bmatrix} 2 & 11 & 3 \\ -2 & -11 & -3 \\ 8 & 35 & 9 \end{bmatrix}$$

The output from the M-file T3_3_6 provides details on this problem. The third power of A is zero and the matrix is nilpotent.

```
>> T3_3_6
A is
        2              11              3
       -2             -11             -3
        8              35              9

A^2 is
        6               6              0
       -6              -6              0
       18              18              0

A^3 is
        0               0              0
        0               0              0
        0               0              0

A is nilpotent
(I-A)^-1 using the sum is
        9              17              3
       -8             -16             -3
       26              53             10

(I-A)^-1 using inv(I-A) is
        9              17              3
       -8             -16             -3
       26              53             10
```

T4. The inverse of $I - A$ can be obtained as a sum under certain conditions.

(a) Use Theorem 3.6.7 to confirm that if

$$A = \begin{bmatrix} 0 & \frac{1}{4} & \frac{1}{8} \\ \frac{1}{4} & \frac{1}{8} & \frac{1}{10} \\ \frac{1}{8} & \frac{1}{10} & \frac{1}{10} \end{bmatrix}$$

then the inverse of $I - A$ can be expressed by the series in Formula (18).

(b) Compute the approximation $(I - A)^{-1} \approx I + A + A^2 + A^3 + \ldots + A^{10}$, and compare it to the inverse of $I - A$ produced directly by your utility. To how many decimal places do the results agree?

Output from the M-file T4_3_6 suggests that the results agree to about four decimal places. The computations are routine except for a MATLAB function `polyvalm` that evaluates the matrix sum of powers. Its explanation and use are given in the M-file.

T5. (CAS) We stated in the text that every strictly triangular matrix is nilpotent. Show that this is true for matrices of size 2×2, 3×3, and 4×4, and make a conjecture about the index of nilpotency of an $n \times n$ strictly triangular matrix. Confirm your conjecture for matrices of size 5×5.

M-file T5_3_6 illustrates this concept using symbolic matrices, and from it you can formulate your own conjecture. All matrices considered for this problem are extracted from the upper left corner of the symbolic matrix A generated in the M-file.

```
A is
[ 0, a, b, c, d]
[ 0, 0, e, f, g]
[ 0, 0, 0, h, r]
[ 0, 0, 0, 0, s]
[ 0, 0, 0, 0, 0]
```

EXERCISE SET 3.7
TECHNOLOGY EXERCISES

T1. (*LU- decomposition*) Technology utilities vary widely on how they handle *LU*-decompositions. For example, some programs perform row interchanges to reduce round off error and hence produce a *PLU*-decomposition, even if an *LU*-decomposition without row interchanges is possible. Determine how your utility handles *LU*-decompositions, and use it to find an *LU*- (or *PLU*)-decomposition of the matrix A in Example 1.

The *PLU*-decomposition of a rectangular matrix in MATLAB is based on the best linear algebra algorithms available. Row exchanges to reduce numerical errors are performed and there is no way to influence that strategy with the standard call `[L,U,P] = lu(A)`. It is important to keep the output list in accordance with the software specification. Otherwise, confusion reigns.

```
>> A = [1 2;3 4]
A =
     1     2
     3     4
>> [L,U,P] =  lu(A); % standard call
>> disp(L) % L is lower triangular
    1.0000         0
    0.3333    1.0000
>> [P,L,U] = lu(A); % permute the output list
>> disp(L) % L is upper triangular and P is lower triangular
    3.0000    4.0000
         0    0.6667
```

There are situations where the *PLU*-decomposition of a large matrix matrix is needed and it is essential that P be the identity matrix. A modified call to `lu` attempts to carry out a *LU*-decomposition without pivots, and that is what is implemented in the function M-file `lupv`. It is successful if it is possible to find such a factorization. A small demonstration indicates its use.

```
>> A = magic(4); % A has 3 linearly dependent columns
>> disp(A)
    16     2     3    13
     5    11    10     8
     9     7     6    12
     4    14    15     1

>> [L,U,P] = lu(A); % standard call and P is not the identity
>> disp(P)
     1     0     0     0
     0     0     0     1
     0     1     0     0
     0     0     1     0

>> [L,U,P,ok] = lupv(A); % ok is 1 if P = I and 0 otherwise
>> disp(ok)
     1

>> disp(L)
    1.0000         0         0         0
    0.3125    1.0000         0         0
    0.5625    0.5663    1.0000         0
    0.2500    1.3012   -3.0000    1.0000

>> disp(U)
   16.0000    2.0000    3.0000   13.0000
         0   10.3750    9.0625    3.9375
         0         0   -0.8193    2.4578
         0         0         0         0
```

```
>> disp(L*U)
   16.0000    2.0000    3.0000   13.0000
    5.0000   11.0000   10.0000    8.0000
    9.0000    7.0000    6.0000   12.0000
    4.0000   14.0000   15.0000    1.0000
```

T1_3_7 applies `lu` and `lupv` to the matrix in Example 1.

T2. (***Back and forward substitution***) *LU*-decomposition breaks up the process of solving linear systems into back and forward substitution. Some utilities have commands for solving linear systems with upper triangular coefficient matrices by back substitution, some have commands for solving linear systems with lower triangular coefficient matrices by forward substitution, and some have commands for using both back and forward substitution to solve linear systems whose coefficient matrix has previously been factored into *LU*- (or *PLU*)-form. Determine whether your utility has any or all of these capabilities, and experiment with them by solving the linear that was considered in Example 1.

There are no special commands because the backslash command x = A\b for solving *Ax* = *b* incorporates strategies for solving upper triangular and lower triangular in an efficient manner.

T3. Use *LU* or *PLU*-decomposition (whichever is more convenient for your utility) to solve the linear systems $A\mathbf{x} = \mathbf{b}_1$, $A\mathbf{x} = \mathbf{b}_2$, and $A\mathbf{x} = \mathbf{b}_3$, where

$$A = \begin{bmatrix} 6 & 2 & -1 & 1 \\ 2 & 7 & 1 & -1 \\ 3 & -1 & 5 & 2 \\ 4 & 3 & 2 & -8 \end{bmatrix}; \quad \mathbf{b}_1 = \begin{bmatrix} 4 \\ 5 \\ 1 \\ 2 \end{bmatrix}, \quad \mathbf{b}_2 = \begin{bmatrix} 0 \\ 4 \\ 2 \\ 3 \end{bmatrix}, \quad \mathbf{b}_3 = \begin{bmatrix} 6 \\ 7 \\ 8 \\ 4 \end{bmatrix}$$

The respective columns in x = A\[b1,b2,b3] are the solutions described in the problem statement. It is equivalent to x = inv(A)*[b1,b2,b3] and much more efficient. T3_3_7 describes the details.

T4. See what happens when you try to find an *LU*-decomposition of a singular matrix.

Absolutely nothing unusual happens. Some of the diagonal elements of U are zero and the M-file T4_3_7 shows the effects when a magic matrix is factored.

EXERCISE SET 3.8
TECHNOLOGY EXERCISES

T1. (***Extracting Submatrices***) Many technology utilities provide methods for extracting rows, columns, and other submatrices of a given matrix. Determine whether your utility has this feature, and, if so, extract the row vectors, column vectors, and the four submatrices of A that are indicated by the partitioning:

$$A = \begin{bmatrix} 1 & 2 & 3 & 4 \\ -3 & 9 & -6 & 12 \\ 7 & 6 & -5 & 2 \\ 0 & 2 & -2 & 3 \end{bmatrix}$$

Submatrices are specified by their row and column indices, and the indexing features in MATLAB are ideal for this type of problem. For instance, A(4,1:3) extracts the lower left matrix in the figure above. The M-file T1_3_8 has extensive comments that show how these submatrices are extracted.

T2. (***Constructing Matrices from Submatrices***) Many technology utilities provide methods for building up new matrices from a set of specified matrices. Determine whether your utility provides for this, and, if so, do the following:

(a) Have your utility construct the matrix A in Exercise T1 from the row vectors of A.
(b) Have your utility construct the matrix A in Exercise T1 from the column vectors of A.

(c) Have your utility construct the matrix A in Exercise T1 from the submatrices A_{11}, A_{12}, A_{21}, and A_{22} indicated by the partitioning in Exercise T1.

Row k of A is selected with $\mathtt{rk} = \mathtt{A(k,:)}$, and vertical concatenation $[\mathtt{r1;r2;r3;r4}]$ reconstructs A from its rows. Horizontal concatenation using commas and the columns of A reconstructs A from its columns. The M-file T2_3_8 illustrates all the requests in this problem.

T3. (***Constructing Block Diagonal Matrices***) Many technology utilities provide methods for constructing block diagonal matrices from a set of specified matrices. Determine whether your utility has this feature, and, if so, use it to construct the block diagonal matrix

$$\begin{bmatrix} A & 0 \\ 0 & A^2 \end{bmatrix}$$

from the matrix in Exercise T1.

This capability exists in MATLAB and the block diagonal matrix is realized as $\mathtt{blkdiag(A, A\char`\^2)}$. There is no practical limit on the number of arguments for $\mathtt{blkdiag}$. Examine and execute the M-file T3_3_8.

T4. Compute the product

$$AB = \begin{bmatrix} 1 & 2 & 3 & 4 \\ 0 & 2 & -1 & 6 \\ 5 & 0 & 3 & 1 \\ -7 & 1 & 3 & 2 \end{bmatrix} \begin{bmatrix} 3 & 3 & -4 & -5 \\ 1 & 0 & 2 & 3 \\ 0 & 1 & 4 & 5 \\ 4 & -4 & -1 & 0 \end{bmatrix}$$

directly and by using the column-row rule of Theorem 3.8.1.

The rule mentioned in Theorem 3.8.1 is an outer product sum for the product. In an informal sense,

$$AB = \sum_{k=1}^{m} A(:,k)B(k,:)$$

where A has m columns. This is displayed in the M-file T4_3_8 to help motivate the concept.

T5. Let A be the 9×9 block diagonal matrix whose successive diagonal blocks are

$$D_1 = \begin{bmatrix} -2 & 3 & 4 \\ 4 & -3 & -3 \\ 4 & -1 & 0 \end{bmatrix}, \quad D_2 = \begin{bmatrix} 0 & 1 & 2 \\ 1 & 0 & 3 \\ 4 & -3 & 8 \end{bmatrix}, \quad D_3 = \begin{bmatrix} -1 & 0 & -5 \\ 1 & 1 & 0 \\ 3 & 2 & 6 \end{bmatrix}$$

Find the inverse of A using Formula (4), and check your result by constructing the matrix A and finding its inverse directly.

Details are given in the M-file T5_3_8 and the matrix A = blkdiag(D1,D2,D3) is normally too large to display using standard display commands. Formatting functions int2str and rats are used to control the size. A visual profile of A is available when the spy command is used, but no picture is shown in this manual. In part, T5_3_8 produces output that looks like a block diagonal matrix.

```
>> T5_3_8
D = blkdiag(D1,D2,D3) is
-2   3   4   0   0   0   0   0   0
 4  -3  -3   0   0   0   0   0   0
 4  -1   0   0   0   0   0   0   0
 0   0   0   0   1   2   0   0   0
 0   0   0   1   0   3   0   0   0
 0   0   0   4  -3   8   0   0   0
 0   0   0   0   0   0  -1   0  -5
 0   0   0   0   0   0   1   1   0
 0   0   0   0   0   0   3   2   6
More output follows.
```

T6. Referring to the matrices in Exercise T5, use Formula (6) to find the inverse of the 6 $\times$ 6 matrix A and check your result by finding the inverse directly.

$$A = \begin{bmatrix} D_1 & D_2 \\ 0 & D_3 \end{bmatrix}$$

80

Once storage is reserved for the inverse

$$A^{-1} = \begin{bmatrix} D_1^{-1} & -D_1^{-1}D_2D_3^{-1} \\ 0 & D_3^{-1} \end{bmatrix}$$

the block matrices in the formula indicate clearly how the fill in the locations. Additional details are given in the M-file T6_3_8.

```
>> T6_3_8
% create A
A = blkdiag(D1,D3);
% Add in the upper right block
A(1:3,4:6) = D2; % note the row and column lists
echo off
A is
-2   3   4   0   1   2
 4  -3  -3   1   0   3
 4  -1   0   4  -3   8
 0   0   0  -1   0  -5
 0   0   0   1   1   0
 0   0   0   3   2   6

The inverse using Formula (6) is
 -3/2      -2      3/2      57     -88      47
   -6      -8       5      194    -301     161
    4       5      -3     -119     185     -99
    0       0       0      -6      10      -5
    0       0       0       6      -9       5
    0       0       0       1      -2       1

The inverse using inv(A) is
 -3/2      -2      3/2      57     -88      47
   -6      -8       5      194    -301     161
    4       5      -3     -119     185     -99
    0       0       0      -6      10      -5
    0       0       0       6      -9       5
    0       0       0       1      -2       1
```

T7. If A is an $n \times n$ matrix, $\mathbf{u}$ and $\mathbf{v}$ are column vectors, and q is a scalar, then the $(n + 1)$ $\times$ $(n + 1)$ matrix

$$B = \begin{bmatrix} A & \mathbf{u} \\ \mathbf{v}^T & q \end{bmatrix}$$

is said to result by **bordering** A with $\mathbf{u}$, $\mathbf{v}$, and q. Border the matrix A in Exercise T1 with

$$\mathbf{u} = \begin{bmatrix} 1 \\ 0 \\ 5 \end{bmatrix}, \quad \mathbf{v} = \begin{bmatrix} 3 \\ 6 \\ -1 \end{bmatrix}, \quad q = 8$$

Matrix concatenation features in MATLAB make this problem easy as the M-file T7_3_8 clearly demonstrates.

T8. In many applications it is important to know the effect on the inverse of an invertible matrix A of changing a single entry in . To explain this, suppose that $A = [a_{ij}]$ is an invertible $n \times n$ matrix whose inverse $A^{-1} = [\gamma_{ij}]$ has column vectors $\mathbf{c}_1, \mathbf{c}_2, \ldots, \mathbf{c}_n$ and row vectors $\mathbf{r}_1, \mathbf{r}_2, \ldots, \mathbf{r}_n$. It can be shown that if a constant λ is added to the entry a_{ij} of A to obtain a matrix B, and if $\ell \neq -1/\gamma_{ji}$, then the matrix B is invertible, and

$$B^{-1} = A^{-1} - \left(\frac{\lambda}{1 + \lambda \gamma_{ji}} \right) \mathbf{c}_i \mathbf{r}_j$$

Consider the matrices

$$A = \begin{bmatrix} 1 & -1 & 1 \\ 0 & 2 & -1 \\ 2 & 3 & 0 \end{bmatrix} \quad \text{and} \quad B = \begin{bmatrix} 1 & -1 & 1 \\ 0 & 2 & -1 \\ 2+\lambda & 3 & 0 \end{bmatrix}$$

(a) Find A^{-1} and B^{-1} for $\lambda = 2$ directly.

(b) Extract the appropriate row and column vectors from A^{-1}, and use them to find B^{-1} using the formula stated above. Confirm that your result is consistent with part (a).

(c) Suppose that an $n \times n$ electronic grid of indicators displays a rectangular array of numbers which forms a matrix $A(t)$ at time t, and suppose that the indicators change one entry at a time at times $t_1, t_2, t_3, \ldots$ in such a way that the matrices $A(t_1), A(t_2), A(t_3), \ldots$ are invertible. Compare the number of flops required to compute the inverse of $A(t_k)$ from the inverse of $A(t_{k-1})$ using the formula stated above as opposed to computing it by row reduction. [***Note:*** Assume that n is large and see Table 1 of Section 3.7.]

Getting the notation straight is the hardest part of this problem, and the extensive commenting in the M-file T8_3_8 helps make the connection between the mathematical symbols and the MATLAB code. The third part is analyzed using the suggested text material. It takes about $2n^2$ flops to calculate the rank one update and add it to the existing inverse. It takes n^2 multiplies to calculate the rank one update and another n^2 additions to produce the final updated inverse. The scalar factor ahead of $c_i r_j$ can be ignored for this calculation. Inverting the matrix from scratch takes about $2n^3$ flops as found in the second to the last entry in the table. One comparison is to take their quotient $2n^3 / 2n^2 = n$, and it shows that the direct approach takes about a thousand times as many flops as the incremental update when $n = 1000$.

MALAB Commands

lu   --- PLU factorization of A

lupv   --- similar to lu but attempts a LU factorization without pivoting

polyvalm --- evaluate a matrix polynomial

ismember --- ismember(A,B) decides if $A \subseteq B$ for matrices A and B

blkdiag --- creates a block diagonal matrices based on the arguments

int2str --- integer formatting function for integer matrices

rats   --- rational formatting function for numerical matrices with nice decimal forms

spy   --- produces a visual profile of a matrix in a figure window

inv   --- inverse of a square matrix

cumsum --- cumulative sum of rows or columns of a matrix

diag   --- create a diagonal matrix and extract a diagonal from a given matrix

triu   --- extract an upper triangular part from a matrix

tril   --- extract a lower triangular part from a matrix

Chapter 4

Determinants

Displaying matrices is common before, during, and after computations. A standard way is to omit the semi-colon or use the `disp` command.

```
>> A = [1 2;3 4]
A =
     1        2
     3        4

>> 2+A
ans =
     3        4
     5        6

>> disp(A)
     1        2
     3        4
```

The first way shows the name of the array and third is a little more compact and does not show the name. A second version relies on the default in MATLAB of assigning the results of a computation to the variable `ans`. There are situations where more information is desirable in a display for reporting purposes. A function with intermediate functionality called `formatA` provides the name and formats the results. It is included with the M-files for this manual and is used extensively in what follows. File operations and formatting data are two of the messier aspects of computing because so many cases must be taken into account. Just as there is no source listing for `disp`, no listing is available for `formatA` at present. Its output is a character array.

```
>> type disp % no listing is available
disp is a built-in function.
>> data = formatA(A) % data is a character matrix with 'A =' on left
data =
     1  2
A = 3  4

>> Af = formatA(A,'The matrix A is')
Af =
                    1  2
The matrix A is = 3  4
```

```
>> S = formatA(A,[]) % an empty second argument formats A
S =
1   2
3   4
```

In the first use, the matrix name is captured by the software and used as part of the output. The second use allows specific text to appear on the left side of the equal sign, and this function M-file is used extensively in what follows. Lastly, A is formatted and nothing is added on the left – just as in disp(A).

EXERCISE SET 4.1
TECHNOLOGY EXERCISES

T1. Compute the determinants in Exercises 23 and 24.

Once a numeric or symbolic matrix A is entered, its determinant is obtained with the command det(A). Examples are presented in the M-file T1_4_1.

T2. Compute the cofactors of the matrix in Example 2, and calculate the determinant of the matrix by a cofactor expansion along the second row and also along the second column.

Calculating a cofactor requires the determinant of a suitable submatrix of A. Code fragments indicate the general idea, and indexing features in MATLAB make it easy to select any submatrix for cofactor evaluation.

```
>> disp(A)

        3       1      -4
        2       5       6
        1       4       8

>> A11 = A(2:3,2:3); % discard first row and column
>> disp(A11);
        5       6
        4       8

% M11 is the minor determinant associated with the position (1,1)
>> M11 = det(A11);
```

```
>> C11 = (-1)^(1+1)*M11; % cofactor evaluation
>> disp(C11)
    16
```

The burden in this problem is to create the appropriate index arrays that determine the submatrix for a particular minor. M-file T2_4_1 shows all the details without any special programming tricks, and a function M-file called `minor` relieves us of most of the drudgery associated with evaluating a minor determinant.

T3. Choose random 3×3 matrices *A* and *B* (possibly using your technology) and then compute $|AB|$ and $|A||B|$. Repeat this process as often as you need to be reasonably certain about the relationship between these quantities. What is the relationship?

Random numbers in the interval [0, 1] are generated with the command rand. A fuller appreciation is possible only if you understand elementary probability. Check `help` `rand` for details. A random 3x3 matrix is generated with the command rand(3,3). After executing the M-file T3_4_1 several times, you can safely assume that $|AB| = |A||B|$ is the requested relationship. More details on random number generation are given in the M-file.

T4. (CAS) Confirm Formula (4) for the determinant of the symbolic 3x3 matrix

$$A = \begin{bmatrix} a_{11} & a_{12} & a_{13} \\ a_{21} & a_{22} & a_{23} \\ a_{31} & a_{32} & a_{33} \end{bmatrix}$$

Symbolic determinants can be evaluated when the Symbolic Toolbox is installed. There is no difficulty once symbolic variables are declared with the command

```
    syms a11 a12 a13 a21 a22 a23 a31 a32 a33
```

Details are given in the M-file T4_4_1.

```
>> T4_4_1
    [a11, a12, a13]
```

```
A = [a21, a22, a23]
    [a31, a22, a33]

det(A) =
a11*a22*a33-a11*a23*a22-a21*a12*a33+a21*a13*a22+a31*a12*a23-a31*a13*a22
This is formula (4) in section 4.1.
```

T5. (CAS) Use the determinant and simplification capabilities of a CAS to show that

$$\begin{vmatrix} \frac{1}{a+x} & \frac{1}{a+y} & 1 \\ \frac{1}{b+x} & \frac{1}{b+y} & 1 \\ \frac{1}{c+x} & \frac{1}{c+y} & 1 \end{vmatrix} = \frac{(a-b)(a-c)(b-c)(x-y)}{(a+x)(b+x)(c+x)(a+y)(b+y)(c+y)}$$

This problem is similar to the last one, and the only point to mention is the way the denominator is displayed in the result. It is correct and unusual, and it is probably due to the simplification algorithm that is invoked when detA = simple(det(A)) is executed. Consult help simple for some details on the simplification algorithm.

```
>> T5_4_1
    [1/(a+x), 1/(a+y), 1]
A = [1/(b+x), 1/(b+y), 1]
    [1/(c+x), 1/(c+y), 1]

det(A) is
(-c+b)*(-c+a)*(-b+a)*(x-y)/(a+x)/(b+y)/(c+y)/(b+x)/(a+y)/(c+x)
```

T6. (CAS) Use the determinant and simplification capabilities of a CAS to show that

$$\begin{vmatrix} a & b & c & d \\ -b & a & d & -c \\ -c & -d & a & b \\ -d & c & -b & a \end{vmatrix} = (a^2+b^2+c^2+d^2)^2$$

Details are in the M-file T6_4_1 and the command simple is used to simplify the raw determinant expression. It is pleasing that the result is identical with the problem statement in contrast with T5.

```
>> T6_4_1
    [ a,  b,  c,  d]
    [-b,  a,  d, -d]
A = [-c, -d,  a,  b]
```

```
        [-d,   c,  -b,    a]
det(A)  =  [(c*d+a^2+d^2+b^2)*(c^2+a^2+d^2+b^2)])
```

Is the command `simple` worth it? Compare the raw output of det(A) with a simplified

version.

```
>> disp(det(A))
a^4+2*a^2*b^2+2*d^2*a^2+c*d*a^2+b^4+2*d^2*b^2+c^2*b^2+
 c*d*b^2+c^2*a^2+c^3*d+c^2*d^2+c*d^3+d^4

>> disp(simple(det(A)))
(c*d+a^2+d^2+b^2)*(c^2+a^2+d^2+b^2)
```

T7. **(CAS)** Find a simple formula for the determinant

$$\begin{vmatrix} (a+b)^2 & c^2 & c^2 \\ a^2 & (b+c)^2 & a^2 \\ b^2 & b^2 & (c+a)^2 \end{vmatrix}$$

A command `simple` in MATLAB simplifies symbolic calculations. Consult the M-file

T7_4_1 for details.

```
>> T7_4_1
       [(a+b)^2,      c^2,       c^2]
A = [     a^2, (b+c)^2,        a^2]
    [     b^2,      b^2, (c+a)^2]


det(A)  =  [2*c*b*a*(a+b+c)^3]
```

T8. The nth-order ***Fibonnaci matrix*** (named for the Italian mathematician (circa 1170 –
1250)) is the $n \times n$ matrix F_n that has 1's on the main diagonal, 1's along the diagonal
immediately above the main diagonal, −1's along the diagonal immediately below the
main diagonal, and zeros everywhere else. Construct the sequence

$$\det(F_1), \ \det(F_2), \ \det(F_3), \ ....., \ \det(F_7)$$

89

Make a conjecture about the relationship between a term in the sequence and its two immediate predecessors, and then use your conjecture to make a guess at $\det(F_8)$. Check your guess by calculating this number.

Several matrices are required, and the command `diag` provides a convenient way to build each one. The next code fragment illustrates how it is done.

```
>> n = 3;
>> F3 = eye(n) + diag(ones(1,n-1),1)+diag(-ones(1,n-1),-1);
>> detF3 = det(F3);
>> disp(formatA(F3));
        1   1   0
F3 =   -1   1   1
        0  -1   1
>> disp(detF3)
detF3 = 3
```

Code below shows how the sub and super diagonal parts of F_n are created using `diag` and `ones`. The second argument of `diag` indicates which diagonal the data is to be placed.

```
>> disp(diag(ones(1,n-1),1)) % superdiagonal with second argument 1
      0      1      0
      0      0      1
      0      0      0
>> disp(diag(-ones(1,n-1),-1)) % subdiagonal with second argument -1
      0      0      0
     -1      0      0
      0     -1      0
```

All the calculations are displayed when T8_4_1 is executed, and the output is too long to include here.

T9. Let A_n be the $n \times n$ matrix that has 2's along the main diagonal, 1's along the diagonals immediately above and below the main diagonal, and zeros everywhere else. Make a conjecture about the relationship between n and $\det(A_n)$.

The operator `diag` is used to build these matrices.

```
>> n = 3;
>> A = diag(2*ones(n,1))+diag(ones(n-1,1),1)+diag(ones(n-1,1),-1);
>> disp(A)
     2     1     0
     1     2     1
     0     1     2
>> det(A)
ans =
     4
```

The number of data items in the first argument of `diag` determines the size of the

resulting matrix.

```
>> diag(1:3)  % 3x3
ans =
     1     0     0
     0     2     0
     0     0     3
>> diag(1:3,1)  % 4x4
ans =
     0     1     0     0
     0     0     2     0
     0     0     0     3
     0     0     0     0
```

It appears that $\det(A) = n+1$, where n is the size of the square matrix A, and the M-file

T9_4_1 gives more details.

EXERCISE SET 4.2
TECHNOLOGY EXERCISES

T1. Find the LU-decomposition of the matrix A, and use it to find det(A) by inspection.

$$A = \begin{bmatrix} -2 & 2 & -4 & -6 \\ -3 & 6 & 3 & -15 \\ 5 & -8 & -1 & 17 \\ 1 & 1 & 11 & 7 \end{bmatrix}$$

Check your result by computing det(A) directly.

The standard tool for determining the LU-decomposition is appropriately called `lu`, and an excerpt from the help file starts to explain how its functionality.

```
>> help lu

 LU      LU factorization.
    [L,U] = LU(X) stores an upper triangular matrix in U and a
    "psychologically lower triangular matrix" (i.e. a product
    of lower triangular and permutation matrices) in L, so
    that X = L*U. X can be rectangular.

    [L,U,P] = LU(X) returns lower triangular matrix L, upper
    triangular matrix U, and permutation matrix P so that
    P*X = L*U.
```

What follows is an extended discussion of this useful command.

```
>> A = [-2 2 -4 -6;-3 6 3 -15;5 -8 -1 1;1 1 11 7];
>> format rat
```

The first use of `lu` generates data that doesn't look familiar and still $A = LU$ holds. Although U is upper triangular, L is not lower triangular and this is explained later.

```
>> [L,U] = lu(A);
>> disp(L) % not a lower triangular matrix
     -2/5             -6/13            -5/18              1
     -3/5              6/13             1                 0
      1                0                0                 0
      1/5              1                0                 0
>> disp(U)
      5               -8               -1                 1
      0               13/5             56/5              34/5
      0                0              -36/13           -228/13
      0                0                0               -22/3
>> disp(L*U) % agrees with A
     -2                2               -4                -6
     -3                6                3               -15
      5               -8               -1                 1
      1                1               11                 7
```

The second use of `lu` produces three matrices and L is lower triangular. The matrix U remains the same. Both P and U are involved in the calculation of det(A) as the code below shows.

```
>> [L,U,P] = lu(A);
>> disp(P)
      0                0                1                 0
      0                0                0                 1
      0                1                0                 0
```

```
        1        0        0        0
>> disp(U)
        5                      -8                     -1                      1
        0                   13/5                   56/5                   34/5
        0                      0                 -36/13                -228/13
        0                      0                      0                  -22/3
>> disp(det(A))
      -264
>> disp(det(U)) % off by a minus sign
       264
>> disp(det(P)) % det(P) is 1 or -1
        -1
```

Next, investigate how L, U, and P interact.

```
>> disp(L) is unit lower triangular
        1                      0                      0                      0
      1/5                      1                      0                      0
     -3/5                   6/13                      1                      0
     -2/5                  -6/13                  -5/18                      1
>> disp(L*U) % is not A because P is not taken into account
        5                     -8                     -1                      1
        1                      1                     11                      7
       -3                      6                      3                    -15
       -2                      2                     -4                     -6
>> disp(P'*L*U) % this does it
       -2                      2                     -4                     -6
       -3                      6                      3                    -15
        5                     -8                     -1                      1
        1                      1                     11                      7
>> disp(A)
       -2                      2                     -4                     -6
       -3                      6                      3                    -15
        5                     -8                     -1                      1
        1                      1                     11                      7
```

It becomes clear in the last two calculations how the matrix L in the first invocation of lu is related to P and L in the second invocation. Since $P^T LU = A$, it must be the case that $P^T L$ from the second invocation agrees with L from the first. A further calculation verifies this conjecture.

```
>> disp(P'*L)
     -2/5                  -6/13                  -5/18                      1
     -3/5                   6/13                      1                      0
        1                      0                      0                      0
      1/5                      1                      0                      0
```

It is the pivoting strategy used in the LU-decomposition that creates a permutation matrix P that is possibly not the identity matrix. There is no easy way to request that a different strategy be used so that P is possibly the identity matrix and still use lu. Indeed, there are simple problems where pivoting is essential. The matrix B comes to mind.

$$B = \begin{bmatrix} 0 & 1 \\ 1 & 0 \end{bmatrix}$$

A function called lupv is available that attempts a LU-decomposition with no row permutations. It is written as a function M-file and it uses special features of lu that are not part of this course.

```
>> B =[0 1;1 0];
>> [L,U,P,ok] = lupv(B);
>> ok
ok =
      0 % pivoting is necessary
>> P'*L*U % returns A
ans =
      0     1
      1     0
```

Pivoting is necessary because is ok is 0. lupv is used in the M-file T1_4_2 to attempt and succeed in obtaining a LU-decomposition without row permutations (P is the identity). No output is included here.

T2. Confirm the formulas in Theorem 4.2.2, for a 5×5 matrix of your choice.

A magic matrix is used to illustrate basic determinant results, and T2_4_2 contains examples of how row swapping and other operations are carried out in MATLAB.

```
>> T2_4_2
Use a magic matrix of size 5
      17    24     1     8    15
      23     5     7    14    16
A =    4     6    13    20    22
      10    12    19    21     3
      11    18    25     2     9

det(A) = 5070000
      *****       *****       *****
Rescale second row of A by 2 and call it A2
      17    24     1     8    15
      46    10    14    28    32
```

```
A2 =    4    6   13   20   22
       10   12   19   21    3
       11   18   25    2    9
det(A2)  =  10140000
2*det(A) =  10140000
        *****        *****        *****
Swap rows 2 and 4 of A and call it A24
       17   24    1    8   15
       10   12   19   21    3
A24 =   4    6   13   20   22
       23    5    7   14   16
       11   18   25    2    9
det(A24) = -5070000
-det(A)  = -5070000
        *****        *****        *****
Add 2 times column 3 to column 5 and call it A235
       17   24    1    8   17
       23    5    7   14   30
A235 =  4    6   13   20   48
       10   12   19   21   41
       11   18   25    2   59
det(A235) = 5070000
det(A)    = 5070000
        *****        *****        *****
```

See help magic for more details on these fascinating matrices.

T3. Let

$$A = \begin{bmatrix} 1 & 3 & 5 & -8 & 9 \\ 11 & 21 & 7 & -3 & 6 \\ -12 & 0 & 3 & 7 & 8 \\ 0 & -3 & 7 & 21 & 3 \\ \varepsilon & -3 & 7 & 21 & 3 \end{bmatrix}$$

   (a) See if you can find a small nonzero value of ε for which your technology utility
       tells you that $\det(A) \neq 0$ and A is not invertible

   (b) Do you think that this contradicts Theorem 4.2.4? Explain.

Getting such a value is not easy in a floating point environment, for the arithmetic
operations used in calculating a determinant introduce roundoff errors that mask the
effect. In most cases, MATLAB reports back that an attempted inverse operation is not
reliable and it produces a best-effort matrix as the inverse.

```
>> C = magic(4);
>> det(C)
ans =
     0

>> Ci = inv(C);
Warning: Matrix is close to singular or badly scaled.
         Results may be inaccurate. RCOND = 1.306145e-017.
>> disp(Ci) % note the large entries in Ci
  1.0e+014 *

     0.9382     2.8147    -2.8147    -0.9382
     2.8147     8.4442    -8.4442    -2.8147
    -2.8147    -8.4442     8.4442     2.8147
    -0.9382    -2.8147     2.8147     0.9382

>> disp(C*Ci)
    1.0000          0    -1.0000    -0.5000
   -0.2500          0          0     0.3750
    0.2500     0.5000          0    -0.2500
    0.1563     0.1250          0     1.2344
```

Similar things happen even when C is close to being singular. The variable eps is built into MATLAB, and it is a small number that is important in numerical linear algebra. Consult help eps for additional details.

```
>> format long e
>> disp(eps)
     2.220446049250313e-016
```

M-file T3_4_2 indicates how a value of ε near 1.5eps has the desired effect.

```
>> T3_4_2
% Find a small entry in the (5,1) position that
% leads to a non-zero determinant but a singular matrix
A = [1 3 5 -8 9;
    11 21 7 -3 6;
    -12 0 3 7 8;
    0 -3 7 21 3;
    0 -3 7 21 3];
A(5,1) = 1.50000000000025*eps;
echo off
A is
  1.0000e+000  3.0000e+000  5.0000e+000 -8.0000e+000  9.0000e+000
  1.1000e+001  2.1000e+001  7.0000e+000 -3.0000e+000  6.0000e+000
 -1.2000e+001            0  3.0000e+000  7.0000e+000  8.0000e+000
            0 -3.0000e+000  7.0000e+000  2.1000e+001  3.0000e+000
  3.3307e-016 -3.0000e+000  7.0000e+000  2.1000e+001  3.0000e+000

Choosing t = 3.3307e-016 leads to det(A) = 3.34e-011
The inverse inv(A) may not exist in a MATLAB environment.
```

96

A*inv(A) leads to nonsense rather than the identity matrix.
```
            1   -0.064174   -0.058826    0    0
            0    0.83984    -0.14681    -2    1
A*inv(A) =  0   -0.039063    0.96419     0    0
            0   -0.20926    -0.19182     1   -1
            0   -0.20926    -0.19182     1    1
```

The entire output is not shown here because inv(A) is considered almost singular and warnings were suppressed.

T4. We know from Exercise 35 that if A is a square matrix, then $\det(AA^T) = \det(A^TA)$. By experimentation, see if you think that the equality always holds if A is not square.

Simple matrices show that the equality doesn't hold when A is not square.

```
>> T4_4_2
A = 1   1
   ***** *****
A*A' = 2
det(A*A') = 2
   ***** *****
A'*A = 1   1
       1   1
det(A'*A) = 0
   ***** *****
```

T5. (CAS) Use a determinant to show that if a, b, c and d are not all zero, then the vectors

$$\mathbf{v}_1 = (a, b, c, d), \quad \mathbf{v}_2 = (-b, a, d, -c), \quad \mathbf{v}_3 = (-c, -d, a, b), \quad \mathbf{v}_4 = (-d, c, -b, a)$$

are linearly independent.

Output from the M-file T5_4_2 tells the whole story.

```
>> T5_4_2
The vectors are the rows of A.
A is
[  a,   b,   c,   d]
[ -b,   a,   d,  -c]
[ -c,  -d,   a,   b]
[ -d,   c,  -b,   a]
The vectors v1, v2, v3, and v4 are linearly independent because
det(A) = (a^2+b^2+d^2+c^2)^2 is not zero unless a, b, c, and d are all
zero.
```

EXERCISE SET 4.3
TECHNOLOGY EXERCISES

T1. Compute the cross products in Example 9.

A cross product command cross works with vectors of size 3, and the first few lines of

T1_4_3 show the essentials.

```
%T1_4_3
% various cross products of u and v
u = [1, 2, -2];
v = [3, 0, 1];
format short

uxv = cross(u,v); % evaluate cross products
vxu = cross(v,u); % order is important
uxu = cross(u,u); % should be the zero vector
```

T2. Compute the adjoint in Example 1, and confirm the computations in Example 2.

Computing the adjoint matrix involves several determinants. A control structure, called a

for loop, is used to automate the process. Formatting the results takes the most space.

```
>> T2_4_3
        3   2  -1
A =  1   6   3
        2  -4   0

det(A)  = 64

                    12    6  -16
cofactors of A =     4    2   16
                    12  -10   16

          12    4   12
adjA =     6    2  -10
         -16   16   16

                                         3/16   1/16    3/16
A^-1 = (1/det(A))*adj(A)   =      3/32   1/32   -5/32
                                        -1/4    1/4     1/4
   and the explicit inverse is

                  3/16   1/16    3/16
inv(A) =      3/32   1/32   -5/32
                 -1/4    1/4     1/4
```

T3. Use Cramer's rule to solve for y without solving for x, z, and w, and check your result by using any method to solve the system:

$$\begin{aligned} 4x+ y+ z+ w &= 6 \\ 3x+7y- z+ w &= 1 \\ 7x+3y-5z+8w &= -3 \\ x+ y+ z+2w &= 3 \end{aligned}$$

Output from T3_4_3 shows that gaussian elimination can exhibit roundoff errors even when the data consists of all integers.

```
>> T3_4_3
        4   1   1   1
        3   7  -1   1
A  =  7   3  -5   8
        1   1   1   2
        6
        1
b  = -3
        3
Replace the second column of A with b and call it A2
        4   6   1   1
        3   1  -1   1
A2 =  7  -3  -5   8
        1   3   1   2
y2  =  det(A2)/det(A)  =  0/-424  =  0

Use gaussian elimination to solve Ax = b.
The computed second component of x is -1.4244e-016
and the discrepancy is due to roundoff error in gaussian elimination.
```

T4. (CAS) Confirm some of the statements in Theorems 4.3.8 and 4.3.9.

This problem involves cross products and the M-file T4_4_3 illustrates results. There is no need to use a CAS because 3-vectors are represented as row vectors.

```
>> T4_4_3
u  =  [1    2    3]
v  =  [1  -1    1]
w  =  [0    1    2]
uxv  =  [5    2  -3]
(u+v)xw  =  [-2  -4    2]
dot(u,uxv)  =  0
dot(v,uxv)  =  0
```

T5. (CAS) Confirm Formula (9) for the fourth and fifth-order Vandermonde determinants.

Part of the output of the M-file T4_4_3 indicates results using symbolic variables.

```
>> T5_4_3
A fourth order Vandermonde matrix V4 is
        [1, x1, x1^2, x1^3]
        [1, x2, x2^2, x2^3]
V4  =   [1, x3, x3^2, x3^3]
        [1, x4, x4^2, x4^3]

det(V4)  =  [(x2-x3)*(x4-x3)*(x4-x2)*(x1-x3)*(x1-x2)*(x1-x4)]
```

The displayed result is identical to the claim made in the text. A function called vander produces a Vandermonde matrix, but it is not designed with symbolic variables in mind. Check doc vander for details.

EXERCISE SET 4.4
TECHNOLOGY EXERCISES

Eigenvalues and eigenvectors are important in linear algebra. There is a fascinating MATLAB demonstration called eigshow that provides an interactive environment for exploring the concept for matrices of size 2. Type eigdshow in the command window and explore.

T1. Find the eigenvalues and corresponding eigenvectors of the matrix in Example 2. If it happens that your eigenvectors are different from those obtained in the example, resolve the discrepancy.

Using [P,D] = eig(A) produces eigenvectors in the columns of P and eigenvectors in the diagonal of D, while evals = eig(A) returns just the eigenvalues in a column vector. Eigenvectors are not unique and be scaled with any nonzero number. In MATLAB, the columns of P are unit vectors and so appear different from the ones usually obtained with hand calculations.

```
>> T1_4_4
      1   3
A = 4  2
```

```
eigenvalues of A = -2   5

Columns of P are corresponding unit eigenvectors
      -0.70711              -0.6
P =   0.70711              -0.8
    ***
Take the eigenvectors from Example 2 and rescale to
create unit vectors. The resulting vectors or their
negatives are the eigenvectors listed above.
****************************************
Solving (A - (-2)I)p = 0 gives an eigenvector corresponding
to lambda = -2. Use solvesys:
*** *** *** *** *** *** ***
The system Ax = b is
|3   3||x1| = |0|
|4   4||x2|   |0|
*** *** *** *** *** *** ***
A computed solution is

|x1| = t1|-1|
|x2|     | 1|
```

Column one of P and the solution reported by solvesys agree within a scaling factor.

T2. Find eigenvectors corresponding to the eigenvalues of the matrix in Example 4.

$$A = \begin{bmatrix} 1/2 & 0 & 0 & 0 \\ -1 & -2/3 & 0 & 0 \\ 7 & 5/8 & 6 & 0 \\ 4/9 & -4 & 3 & 6 \end{bmatrix}$$

Eigenvalues are 6, 6, -2/3, and 1/2. The eigenspace for the double eigenvalue 6 is one dimensional, and this signals a geometric multiplicity of one. The M-file T2_4_4 determines eigenvectors in two ways. The columns of the matrix P are eigenvectors of A when [P,D] = eig(A) is executed. This problem illustrates an important feature of the eig command. Two columns of P are identical even though $AP = PD$ and P is not invertible. Another way to obtain eigenvectors is to separately ask for the null space of $A - \lambda I$ when λ is one of the distinct eigenvalues. null(A -0.5*eye(4),'r') produces an eigenvector for the eigenvalue 1/2, and the other eigenvectors are calculated in a similar fashion.

T3. Define an nth-order **checkerboard matrix** C_n to be a matrix that has a 1 in upper left corner and alternates between 1 and 0 along rows and columns (see accompanying figure for an example).

  (a) Find the eigenvalues of C_1, C_2 , C_3, C_4, C_5 and make a conjecture about the eigenvalues of C_6. Check your conjecture by finding the eigenvalues of C_6.

  (b) In general, what can you say about the eigenvalues of C_n ?

1	0	1	0
0	1	0	1
1	0	1	0
0	1	0	1

It is important to write out a few more cases of the checkerboard matrices. Start with the largest one because it holds the clue for producing any of them.

```
>> C = repmat(eye(2),4,4); % This is really C8
>> disp(int2str(C))
1   0   1   0   1   0   1   0
0   1   0   1   0   1   0   1
1   0   1   0   1   0   1   0
0   1   0   1   0   1   0   1
1   0   1   0   1   0   1   0
0   1   0   1   0   1   0   1
1   0   1   0   1   0   1   0
0   1   0   1   0   1   0   1
```

Start in the upper left corner and grab the first n rows and n columns – it is C_n. Matrix C is C_8 and the indexing features in MATLAB make it simple to generate this matrix. The function `repmat` takes a 2x2 identity matrix and replicates the pattern 4 times across and 4 times down to deliver C. Look into C and see the eight identity matrices of size two. Patterns are an important part of making linear algebra easier with MATLAB. Powerful primitives in MATLAB provide novel approaches for matrix creation that are not possible if you concentrate only on the individual matrix elements. M-file T3_4_4 provides all the data needed to answer the questions.

T4. Confirm the statement in Theorem 4.4.6 for the matrix in Example 2 with $k = 2, 3, 4,$ and 5.

Output associated with the commented M-file for this problem demonstrates the general result concerning the eigenvalues of powers of a matrix. Eigenvalues of A^5 are the fifth power of the eigenvalues of A, with similar results for other powers.

```
>> T4_4_4
        1    3
A =     4    2
eig(A)  =   -2    5
    *****        *****
        13    9
A^2 =   12    16
eig(A^2)  =   4    25
    *****        *****
        1321    1353
A^5 =   1804    1772
eig(A^5)  =   -32    3125
    *****        *****
```

T5. (CAS)

(a) Use the command for finding characteristic polynomials to find the characteristic polynomial of the matrix in Example 3, and check the result by using a determinant operation.

(b) Find the exact eigenvalues by solving the characteristic equation.

Finding the characteristic polynomial for a matrix is a delicate process except when the entries are small integers as is the case here. Although M-file T5_4_4 automates much of the work requested in this problem, these code fragments illustrate the main points.

```
>> A = [0 -1 0;0 0 1;-4 -17 8]; % matrix
>> syms t % symbolic variable
>> cpoly = det(A - t*eye(3)); % characteristic polynomial
>> disp(cpoly) % note the unusual output format

8*t^2-t^3-17*t+4
```

```
>> theroots = solve(cpoly); % toolbox call to get exact symbolic roots
>> disp(theroots)
[          4]
[ 2+3^(1/2)]
[ 2-3^(1/2)]
>> disp(factor(cpoly)) % factor makes it easy to verify the roots

-(t-4)*(t^2-4*t+1)
```

A symbolic calculation is used to evaluate the determinant. Getting the exact roots requires a call to the `solve` routine, and the command `factor` puts the characteristic polynomial in a form that is easy to interpret.

T6. (CAS) Obtain the formulas in Exercise 30.

The command `syms` creates symbolic variables, and `collect` organizes the characteristic polynomial by powers. As an experiment, remove the collect command and observe the resulting symbolic output.

```
%T6_4_4
% Obtain formula for eigenvalues of a symbolic matrix
syms a b c d t % symbols for entries
A = [a b;c d];

cpoly = collect(det(A - t*eye(2))); % polynomial in t
sol = simple(solve(cpoly)); % obtain roots and simplify
. . .display code is not shown here
```

Results are given in a form that leaves something to be desired. As an experiment, use `simple`, `factor`, `simplify`, or any other command you can think of to obtain a cleaner expression for the roots.

```
>> T6_4_4
     [a, b]
A =  [c, d]

p(t) = det(A - t*eye(2)) = [t^2+(-a-d)*t+a*d-b*c]
The roots are given in symbolic form.
        [1/2*a+1/2*d+1/2*(a^2-2*a*d+d^2+4*b*c)^(1/2)]
Roots = [1/2*a+1/2*d-1/2*(a^2-2*a*d+d^2+4*b*c)^(1/2)]
```

T7. Graph the characteristic polynomial of the matrix

$$A = \begin{bmatrix} 1 & 1 & 2 & 1 & 1 \\ 1 & 2 & 3 & 2 & 1 \\ 2 & 3 & 1 & 2 & 1 \\ 1 & 2 & 2 & 3 & 1 \\ 1 & 1 & 1 & 1 & 7 \end{bmatrix}$$

and estimate the roots of the characteristic equation. Compare your estimates to the eigenvalues produced directly by your utility.

The graph below looks useful until the scaling is examined, for the vertical range precludes any reasonable estimates of the roots. There are two distinct roots between 0 and 2, but the graph makes it appear there might even be a double root. Computed roots and the graph come from the M-file.

```
>> T7_4_4
The eigenvalues obtained by using eig(A):
         -1.8984
          0.31262
eig(A) =  1.0712
          5.2807
          9.2339
```

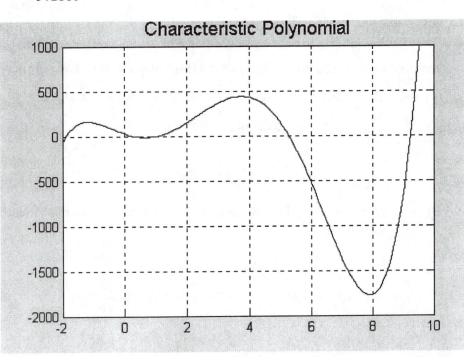

Characteristic Polynomial

The lesson offered in this problem is that polynomial root finding using graphics is not a reliable strategy.

T8. Select some pairs of 3×3 matrices *A* and *B*, and compute the eigenvalues of *AB* and *BA*. Make an educated guess about the relationship between the eigenvalues of *AB* and *BA* in general.

Several pairs of matrices are selected and the eigenvalues of the products suggested are determined and displayed in the M-file T8_4_4. Conclusions are offered in the output. Complex eigenvalues can occur even when the matrices have real entries. Special matrices used in this problem are a magic matrix and the inverse of a Hilbert matrix.

```
>> disp(magic(3))
     8     1     6
     3     5     7
     4     9     2
>> disp(invhilb(3))
     9   -36    30
   -36   192  -180
    30  -180   180
```

They are a few of several special matrices provided as part of MATLAB. Add up the rows and columns of magic(3) to see a possible origin of the name. The number 15 comes up every time. Both types of matrices come in various sizes. Consult help magic and help invhilb for more details.

MATLAB Summary

lupv --- attempt a factorization A = LU without pivots. Included with manual files.

magic --- magic matrices

eigshow --- MATLAB demo for exploring eigenvalue concepts

repmat --- reproduces a given matrix based on replication information

det --- det(A) calculates the determinant of A

formatA --- formats a matrix with optional names

eig --- eig(A) gives eigenvalues and [P,D] = eig(A) gives corresponding eigenvectors in P

collect --- symbolic routine to collect terms in algebraic expressions

simple --- simplifies symbolic expressions

Chapter 5
Matrix Models

EXERCISE SET 5.1
TECHNOLOGY EXERCISES

T1. By calculating sufficiently high powers of P, confirm the result in part (b) of Exercise D5 for the matrix

$$P = \begin{bmatrix} 0.2 & 0.4 & 0.5 \\ 0.1 & 0.3 & 0.1 \\ 0.7 & 0.3 & 0.4 \end{bmatrix}$$

Powers of P become constant after a while, and part of the output of M-file T1_5_1 shows the limiting matrix for the powers.

```
          3/8   3/8   3/8
P^30 =    1/8   1/8   1/8
          1/2   1/2   1/2
```

The columns are identical for all practical purposes.

T2. Each night a guard patrols an art gallery with seven rooms connected by corridors, as shown in the accompanying figure. The guard spends 10 minutes in a room and then moves to a neighboring room that is chosen at random, each possible choice being equally likely.

(a) Find the 7×7 transition matrix for the surveillance pattern.

(b) Assuming that the guard (or a replacement) follows the surveillance pattern indefinitely, what proportion of time does the guard spend in each room?

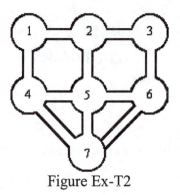

Figure Ex-T2

The hardest part of this problem is to get the data correct for the transition matrix T. The strategy employed here is to first find the adjacency matrix for the graph, taking into account the specification that the guard must move to a different room every ten minutes. That matrix is

$$
Adj = \begin{array}{ccccccc}
0 & 1 & 0 & 1 & 0 & 0 & 0 \\
1 & 0 & 1 & 0 & 1 & 0 & 0 \\
0 & 1 & 0 & 0 & 0 & 1 & 0 \\
1 & 0 & 0 & 0 & 1 & 0 & 1 \\
0 & 1 & 0 & 1 & 0 & 1 & 1 \\
0 & 0 & 1 & 0 & 1 & 0 & 1 \\
0 & 0 & 0 & 1 & 1 & 1 & 0
\end{array}
$$

Rooms are labeled according to the matrix entry. Adj (1,2) is 1 because the guard might visit room 2 after standing guard in room 1. The matrix is symmetric and is zero on the diagonal. The sum of each column is used as a scaling factor for that column so as to introduce a relative frequency of visitation and ensure that the columns of the transition matrix T sum to one. Some command window output of the M-file T2_5_1 illustrates the main points.

$$
T = \begin{array}{ccccccc}
0 & 1/3 & 0 & 1/3 & 0 & 0 & 0 \\
1/2 & 0 & 1/2 & 0 & 1/4 & 0 & 0 \\
0 & 1/3 & 0 & 0 & 0 & 1/3 & 0 \\
1/2 & 0 & 0 & 0 & 1/4 & 0 & 1/3 \\
0 & 1/3 & 0 & 1/3 & 0 & 1/3 & 1/3 \\
0 & 0 & 1/2 & 0 & 1/4 & 0 & 1/3 \\
0 & 0 & 0 & 1/3 & 1/4 & 1/3 & 0
\end{array}
$$

$$
T^{100} = \begin{array}{ccccccc}
0.1 & 0.1 & 0.1 & 0.1 & 0.1 & 0.1 & 0.1 \\
0.15 & 0.15 & 0.15 & 0.15 & 0.15 & 0.15 & 0.15 \\
0.1 & 0.1 & 0.1 & 0.1 & 0.1 & 0.1 & 0.1 \\
0.15 & 0.15 & 0.15 & 0.15 & 0.15 & 0.15 & 0.15 \\
0.2 & 0.2 & 0.2 & 0.2 & 0.2 & 0.2 & 0.2 \\
0.15 & 0.15 & 0.15 & 0.15 & 0.15 & 0.15 & 0.15 \\
0.15 & 0.15 & 0.15 & 0.15 & 0.15 & 0.15 & 0.15
\end{array}
$$

Take any column of T100 to obtain the proportion of time
spent in each of the rooms.

```
room    proportion
1          0.1
2          0.15
3          0.1
4          0.15
5          0.2
6          0.15
7          0.15
```

T3. Acme trucking rents trucks in New York, Boston, and Chicago, and the trucks are
returned to those cities in accordance with the accompanying table. Determine the
distribution of the trucks over the long run.

Trucks rented at			Trucks Returned To
New York	Boston	Chicago	
0.721	0.05	0.211	New York
0.122	0.92	0.095	Boston
0.157	0.03	0.694	Chicago

Table Ex-T3

The transition matrix is given, and the only new feature is the use of the function
strvcat to concatenate several character strings and make a table of cities.

```
>> T3_5_1
        721/1000    1/20     211/1000
T =      61/500    23/25     19/200
        157/1000   3/100     347/500


          0.24007   0.24007   0.24007
T^100 =   0.5799    0.5799    0.5799
          0.18003   0.18003   0.18003
```

The distribution of trucks over the
long haul is in the table.

```
New York   0.24007
Boston     0.5799
Chicago    0.18003
```

The mixed formatting on T and its powers is a consequence of the logic in `formatA`.

EXERCISE SET 5.2
TECHNOLOGY EXERCISES

T1. Suppose that the consumption matrix for an open economy is

$$C = \begin{bmatrix} .29 & .05 & .04 & .01 \\ .02 & .31 & .01 & .03 \\ .04 & .02 & .44 & .01 \\ .01 & .03 & .04 & .32 \end{bmatrix}$$

(a) Confirm that the economy is productive, and then show by direct computation that $(I - C)^{-1}$ has positive entries.

(b) Use matrix inversion or row reduction to find the production vector $\mathbf{x}$ that satisfies the demand

$$\mathbf{d} = \begin{bmatrix} 200 \\ 100 \\ 350 \\ 275 \end{bmatrix}$$

Once the data is entered, most of the effort is in formatting the results as shown in T1_5_2.

```
>> T1_5_2
                            29/100     1/20      1/25     1/100
                            1/50      31/100    1/100     3/100
consumption matrix C =      1/25       1/50     11/25     1/100
                            1/100      3/100     1/25     8/25

The economy is productive because the entries
of (I - C)^-1 are nonnegative.
                  1.4178      0.10696     0.10512    0.027114
                  0.043847    1.4563      0.033808   0.065389
(I - C)^-1 =      0.10335     0.060889    1.7964     0.030623
                  0.028863    0.069402    0.10871    1.4757

With a demand vector d
      200
      100
d = 350
```

```
      275
the corresponding production vector x = inv(I-C)*d is
      338.4968
      184.2118
x =   663.9104
      456.5702
```

T2. The Leontief equation $\mathbf{x} - C\mathbf{x} = \mathbf{d}$ can be rewritten as

$$\mathbf{x} = C\mathbf{x} + \mathbf{d}$$

and solved approximately by substituting an arbitrary initial approximation $\mathbf{x}_0$ into the right side of this equation and using the resulting vector $\mathbf{x}_1 = C\mathbf{x}_0 + \mathbf{d}$ as a new (and often better) approximation to the solution. By repeating this process, you can generate a succession of approximations $\mathbf{x}_1, \mathbf{x}_2, \mathbf{x}_3, \ldots, \mathbf{x}_k, \ldots$recursively from the relationship $\mathbf{x}_k = C\mathbf{x}_{k-1} + \mathbf{d}$. We will see in the next section that this sequence converges to the exact solution under fairly general conditions. Take $\mathbf{x}_0 = \mathbf{0}$, and use this method to generate a succession of ten approximations, $\mathbf{x}_1, \mathbf{x}_2, \ldots, \mathbf{x}_{10}$, to the solution of the problem in Exercise T1. Compare $\mathbf{x}_{10}$ to the result obtained in that exercise.

M-file output gives the details, and you can make a visual comparison between the approximations and the results of T1. The M-file generates a matrix called xiter that contains the intermediate results as column vectors.

```
>> T2_5_2
                         29/100   1/20     1/25     1/100
                         1/50     31/100   1/100    3/100
consumption matrix C =   1/25     1/50     11/25    1/100
                         1/100    3/100    1/25     8/25

With a demand vector d
      200
      100
d =   350
      275
the corresponding production vector x10 after 10 iterations
using the scheme xnew = C*xcurrent + d is compared with the result x
obtained by using gaussian elimination in (I-C)x = d.

     x           x10
   338.4968    338.3995
   184.2118    184.1540
   663.9104    663.6033
   456.5702    456.4602
```

T3. Consider an open economy described by the following table:

Purchased From

	Agriculture	Manufacturing	Trade	Services	Energy
Agriculture	0.27	0.39	0.03	0.02	0.23
Manufacturing	0.15	0.15	0.10	0.01	0.22
Trade	0.06	0.07	0.36	0.15	0.35
Services	0.27	0.08	0.07	0.41	0.09
Energy	0.23	0.19	0.36	0.24	0.10

(a) Show that the sectors are all profitable.

(b) Find the production levels that will satisfy the following demand by the open sector (units in millions of dollars):

Agriculture: $1.2   Manufacturing: $3.4   Trade: $2.7
Services: $4.3   Energy: $2.9

(c) If the demand for services doubles from the level in part (b), which sector will be affected the most? Explain your reasoning.

After entering the data for the consumption matrix, the numbers quoted in part (b) are used to create a demand vector. The remainder of the problem follows the general plan laid out in previous problems.

```
>> T3_5_2
With a demand vector d
        6/5
        17/5
d =    27/10
       43/10
       29/10
the corresponding production vector x = inv(I-C)*d is
       33.485
       27.3598
x =    44.2852
       38.5229
       45.5423
```

114

After doubling the fourth entry of d so that services
is doubled, the new production vector xmod is
```
         40.9845
         33.0919
xmod =   55.2503
         53.0585
         56.9312
```
The component most affected is determined by comparing
both production vectors x and xmod, and then determining
which component changed the most in a relative sense using
the maximum absolute deviation.

The economy component 'services' changed the most when
the services component of demand was doubled.
The change in percent is 37.7324.

It is possible to determine which component of the economy is affected most by looking

at the fourth column of $(I-C)^{-1}$. The fourth component of column four is the largest

value in that column. It corresponds to services and this is the column calculated in the

M-file.

```
              3.5089    2.519   2.1731   1.7441    2.532
              1.7987   2.7014   1.7372   1.3331   1.9289
inv(I-C)  =   2.8646   2.6351   4.2304    2.55    3.2763
              2.6675    2.27    2.221    3.3804   2.4384
              3.1336   2.8734   3.2065   2.6486   4.1262
```

Why? The modified demand vector dmod can be thought of as $d+d(4)e_4$, where e_4 is

the fourth column of a 4x4 identity matrix. Equations $(I-C)x=d$,

$(I-C)x_{mod} = d+d(4)e_4$ can be combined to yield $x_{mod} - x = d(4)(I-C)^{-1}e_4$. But

$(I-C)^{-1}e_4$ is the fourth column of the inverse and this is the clue as to why the column

four is so important in answering this part of the question.

EXERCISE SET 5.3
TECHNOLOGY EXERCISES

In Exercises T1 and T2, approximate the solutions using Jacobi iteration, starting with

$x = 0$ and continuing until two successive iterations agree to 3 decimal places. Do the

same using Gauss-Seidel iteration and compare the number of iterations required by each

method.

T1. The system of Exercise 3.

Both iteration methods, Jacobi and Gauss-Seidel, are to be used for the system $Ax = b$.

```
>> T1_5_3
       10  1   2
A =     1 10  -1
        2  1  10


         3
b =     3/2
        -9


       10  0   0
D =     0 10   0
        0  0  10


Jacobi iteration is xnew = inv(D)*(D-A)*xcurrent + inv(D)*b

The approximate error at iteration 1 is 10^0
The approximate error at iteration 2 is 10^-1
The approximate error at iteration 3 is 10^-2
The approximate error at iteration 4 is 10^-2
The approximate error at iteration 5 is 10^-3
Termination after 5 iterations

          0.49968
x5 =      0.00024
         -0.99984
   ***   ***   ***   ***   ***   ***   ***   ***   ***
Gauss-Seidel iteration is xnew = inv(D-L)*U*xcurrent + inv(D-L)*b

The approximate error at iteration 1 is 10^0
The approximate error at iteration 2 is 10^-1
The approximate error at iteration 3 is 10^-2
The approximate error at iteration 4 is 10^-3
Termination after 4 iterations

           0.49992
x4 =  2.4825e-005
          -0.99999


   *** Iteration Summary ***

  Jacobi    Gauss-Seidel       A\b
     0.4997      0.4999      0.5000
     0.0002      0.0000           0
    -0.9998     -1.0000     -1.0000
```

The summary shows that Gauss-Seidel achieved the termination condition earlier than Jacobi.

T2. The system of Exercise 4.

This problem is similar to the last one and no discussion of T2_5_3 is given.

T3. Consider the linear system
$$2x_1 - 4x_2 + 7x_3 = 8$$
$$-2x_1 + 5x_2 + 2x_3 = 0$$
$$4x_1 + x_2 + x_3 = 5$$

(a) The coefficient matrix of the system is not diagonally dominant, so Theorem 5.3.1 does not guarantee convergence of either Jacobi iteration or Gauss-Seidel iteration. Compute the first five Gauss-Seidel iterates to illustrate a lack of convergence.

(b) Reorder the equations to produce a linear system with a diagonally dominant coefficient matrix, and compute the first five Gauss-Seidel iterates to illustrate convergence.

The lack of convergence for the Gauss-Seidel algorithm for the original system is obvious.

```
>> T3_5_3
   ***    ***    ***    ***    ***    ***    ***    ***    ***
Gauss-Seidel iteration is xnew = inv(D-L)*U*xcurrent + inv(D-L)*b

        2 -4   7
A =   -2   5   2
        4   1   1

        8
b =     0
        5

     *** Iteration Summary ***

        x1          x2          x3          x4          x5          A\b
    4.0000e+0   5.1300e+1   8.4528e+2   1.4177e+4   2.3802e+5   1.0160e+0
    1.6000e+0   2.5560e+1   4.2842e+2   7.1925e+3   1.2077e+5   5.3191e-2
   -1.2600e+1  -2.2576e+2  -3.8045e+3  -6.3894e+4  -1.0728e+6   8.8298e-1
```

```
   ***    ***    ***    ***    ***    ***    ***    ***    ***
Solve Ax = b using Gauss-Seidel iteration.
Gauss-Seidel iteration is xnew = inv(D-L)*U*xcurrent + inv(D-L)*b

      7 -4  2
A =   2  5 -2
      1  1  4

      8
b =   0
      5
    *** Iteration Summary ***

       x1          x2          x3          x4          x5          A\b
      1.1429      0.5735      0.9566      0.8796      0.8797      0.8830
     -0.4571      0.2020      0.0398      0.0485      0.0553      0.0532
      1.0786      1.0561      1.0009      1.0180      1.0162      1.0160
```

Gauss-Seidel is converging for the reordered system.

T6. **Heat** is energy that flows from a point with a higher temperature to a point with a
lower temperature. This energy flow causes a temperature decrease at the point with
higher temperature and a temperature increase at the point with lower temperature until
the temperatures at the two points are the same and the energy flow stops; the
temperatures are then said to have reached a **steady state**. We will consider the problem
of approximating the steady-state temperature at the interior points of a rectangular metal
plate whose edges are kept at fixed temperatures. Methods for finding steady-state
temperatures at all points of a plate generally require calculus, but if we limit the problem
to finding the steady-state temperature at a finite number of points, then we can use linear
algebra. The idea is to overlay the plate with a rectangular grid and approximate the
steady-state temperatures at the interior grid points. For example, the figure below shows
a rectangular plate with fixed temperatures of 0°C, 0°C, 4°C, and 16°C on the edges and
unknown steady-state temperatures $t_1, t_2, \ldots, t_9$ at nine interior grid points. Our approach
will use the **discrete averaging** model from thermodynamics, which states that steady-
state temperature at an interior grid point is approximately the average of the
temperatures at the four adjacent grid points. Thus, for example, the steady-state
temperatures t_5 and t_1 are

$$t_5 = \tfrac{1}{4}\left(t_4 + t_2 + t_6 + t_8\right) \quad \text{and} \quad t_1 = \tfrac{1}{4}\left(0 + 0 + t_2 + t_4\right)$$

(a) Rewrite Write out all nine discrete averaging equations for the steady-state temperatures t_1, t_2, ..., t_9.

(b) Rewrite the equations of part (a) in the matrix form $\mathbf{t} = B\mathbf{t} + \mathbf{c}$, where $\mathbf{t}$ is the column vector of unknown steady-state vectors, B is a 9×9 matrix, and $\mathbf{c}$ is 9×1 column vector of constants

(c) The form of the equation $\mathbf{t} = B\mathbf{t} + \mathbf{c}$ is the same as Equation (1), so Equation (2) suggests that it can be solved iteratively using the recurrence relation $\mathbf{t}_{k+1} = B\mathbf{t}_k + \mathbf{c}$, as with Jacobi iteration. Use this method to approximate the steady-state temperatures, starting with $t_0 = 0, t_2 = 0, \cdots, t_9 = 0$, and using 10 iterations. [*Comment*: This is a good example of how sparse matrices arise – each unknown depends on only a few of the other unknowns, so the resulting matrix has many zeros.]

(d) Approximate the steady-state temperatures using as many iterations as are required until the difference between two successive iterations is less than 0.001 in each entry.

There are sophisticated ways to address the heat conduction problem, and the approach presented is not one of them. What is given here is an approach that uses MATLAB to build patterns needed to understand the nature of the problem as explained in the text. Codes are written with enough flexibility to encourage experimentation as part of a class project.

Writing out a few of the patterns is instructive, but it is easy to make an error and there is the ongoing problem of organizing the data so that the results can be calculated and presented. Almost the first thing that comes to mind is to question what happens if more points are requested or the boundary conditions are changed? That concern is addressed in T6sym_6_5 and T6num_6_5. Starting with the symbolic calculations, you can see the general structure of the problem. After that, run the numerical version because of performance issues related to extensive symbolic calculations. The code in T6num_6_5 is organized so that you can pick the number of points and see a surface presentation of the resulting approximation. You can also change the number of iterations for the iterative approach because twenty is not enough. A careful person can even change the boundary conditions and see other temperature profiles. Matrices associated with this problem are large, contain many zeros, and are referred to as a sparse matrix. Presenting sparse matrices is a problem, and MATLAB designers provide a function called spy that gives a profile of a sparse matrix. Calling spy(A) causes a figure window to appear with dots for locations where non-zero data is present. In schematic form,

$$A = \begin{bmatrix} 2 & 0 & 1 \\ 0 & 0 & 2 \\ 1 & -2 & 5 \end{bmatrix} \rightarrow \text{spy}(A) \rightarrow \begin{bmatrix} \bullet & & \bullet \\ & & \bullet \\ \bullet & \bullet & \bullet \end{bmatrix}$$

This function is used in both M-files to provide a profile of the sparse matrix B that is part of the iteration process. Read on and read the M-files.

Problems with a geometric interpretation usually have an inherent addressing scheme that is easy on the eyes. When the temperature is thought of as a function $t(x,y)$, it is appropriate to consider a discretization process determined by the grid shown in the picture. For some increment h that turns out to be unimportant at this time, set $x_i = ih$, $y_j = jh$, and $t_{ij} = t(x_i, y_j)$, $i, j = 0, \ldots N$. Boundary values correspond to a subscript of 0 or N, and $N = 4$ for the problem described here. The goal is to get an approximation to the temperature at the grid points using the discrete averaging strategy. The temperature labeling strategy given in the problem statement can be improved if the above addressing scheme is used. It leads to an automatic procedure for generating B and c, and this is of some interest for anyone who attempts the problem by hand. Symbolic toolbox calculations are used first to get a feeling for the patterns in this problem, and the T6sym_5_3 provides the details.

```
>> T6sym_5_3
```

The physical view of the temperatures uses
the first coordinate as an x index and the
second as a y index for the body. The physical
temperature matrix is

```
       [t04, t14, t24, t34, t44]
       [t03, t13, t23, t33, t43]
Tp =   [t02, t12, t22, t32, t42]
       [t01, t11, t21, t31, t41]
       [t00, t10, t20, t30, t40]
```

Rotate 90 degrees clockwise to get an indexing
scheme that looks like a matrix with index origin 0.

```
         [t00, t01, t02, t03, t04]
         [t10, t11, t12, t13, t14]
Trot =   [t20, t21, t22, t23, t24]
         [t30, t31, t32, t33, t34]
         [t40, t41, t42, t43, t44]
```

To coordinate it with MATLAB graphics, look at a flip
of the physical temperature matrix.

```
      [t00, t10, t20, t30, t40]
      [t01, t11, t21, t31, t41]
T =   [t02, t12, t22, t32, t42]
      [t03, t13, t23, t33, t43]
      [t04, t14, t24, t34, t44]
```

The interior temperatures to be estimated are

```
                              [t11, t21, t31]
Interior Temperatures = Tint = [t12, t22, t32]
                              [t13, t23, t33]
```

The transposed vector of known boundary temperatures is
transpose(Tb) = [t10,t20,t30,t14,t24,t34,t01,t02,t03,t41,t42,t43]
The order inherent in this vector is found
in the physical temperature matrix Tp -- bottom, top, left, right.
Iteration commences once values are assigned to these variables.
*** *** *** *** *** *** *** *** ***
The interior temperatures are the solution of t = B*t + c
where the transposed vector x is given symbolically as
transpose(Tvec) = [t11, t12, t13, t21, t22, t23, t31, t32, t33]

```
        0  1  0  1  0  0  0  0  0
        1  0  1  0  1  0  0  0  0
        0  1  0  0  0  1  0  0  0
        1  0  0  0  1  0  1  0  0
4*B =   0  1  0  1  0  1  0  1  0
        0  0  1  0  1  0  0  0  1
        0  0  0  1  0  0  0  1  0
        0  0  0  0  1  0  1  0  1
        0  0  0  0  0  1  0  1  0
```

In the iteration t = B*t + c

```
   [1/4*t10+1/4*t01]
   [        1/4*t02]
   [1/4*t14+1/4*t03]
```

```
      [         1/4*t20]
C = [              0]
      [         1/4*t24]
      [1/4*t30+1/4*t41]
      [         1/4*t42]
      [1/4*t34+1/4*t43]
```

This concludes the symbolic analysis of the problem.

To get some idea of how B could be created with an automatic procedure, look at the first node in position (1,1). The averaging principle says that

$$t_{11} = \frac{1}{4}(t_{10} + t_{01} + t_{21} + t_{12})$$

Ignore the factor ¼ for now. A transposed solution vector is compared with this expression to obtain the coefficients of the first row of $4B$. It is a vector of zeros and ones. There is a one in a position if the corresponding symbolic component is in the expression, and zero otherwise. The second row in the next matrix is the first row of $4B$.

$$
\begin{array}{ccccccccc}
t_{11} & t_{21} & t_{31} & t_{12} & t_{22} & t_{32} & t_{13} & t_{23} & t_{33} \\
0 & 1 & 0 & 1 & 0 & 0 & 0 & 0 & 0
\end{array}
$$

Set ideas are exploited and this array can be realized in two steps with

```
>> disp(transpose(Tvec))
[ t11, t12, t13, t21, t22, t23, t31, t32, t33]

>> disp(ismembersym(Tvec,[t21,t12,t10,t01])')
     0     1     0     1     0     0     0     0     0
```

The function ismembersym answers the question of which elements of Tvec are found in the second argument? This function M-file is included with the other M-files for the chapter, and it is modeled on a similar MATLAB function that apparently doesn't accept symbolic arguments. The dot product of these two row vectors gives the first part of the expression

$$4t_{11} = (t_{21} + t_{12}) + (t_{10} + t_{01})$$

A similar vector for the boundary terms gives the first row of a matrix C that produces c when multiplied into the boundary value vector mentioned in the M-file output. The last display from that M-file lists c, and the first entry is $\frac{1}{4}(t_{10} + t_{01})$, as anticipated. Repeat

```
```

the process for each interior point in the order listed in a symbolic vector Tvec of symbolic discrete temperatures. For example, averaging at the next grid point is

$$4t_{12} = \left(t_{22} + t_{13} + t_{11}\right) + t_{02}$$

This is why the second row of $4B$ has three ones in it. The resulting matrix $4B$ has many zeros and it is usually called sparse.

With so much structure available, we can automate the process for a grid of arbitrary density. The connections between the physical temperature matrix and its flipped form provide a way to visualize the results using basic MATLAB commands. Details are in the M-file T6num_5_3, where a strictly numeric approach is taken. There is still a way to use the same basic strategy when it is realized that associated with each matrix is a vector and an integer that points to the data.

```
>> A = [1 2;3 4];
>> disp(A)
     1     2
     3     4
>> disp(A(:))
     1
     3
     2
     4
>> disp([[(1:4)',A(:)])
     1     1
     2     3
     3     2
     4     4
```

The operation A (:) stacks the columns of A, and the first column is a pointer into the column vector A (:). Knowing an integer in the first column makes it possible to extract any element of A using A (:). Conversely, a location in the matrix A points to a definite element in A (:). Symbols are replaced by integers because the main reason for providing a symbolic approach is to motivate a numerical approach in T6num_5_3. Aspects of the iteration process and other featues of this problem are carried out in that file.

```
>> T6num_5_3
A spy plot of the matrix B for the iteration tnew = Bt + c
shows the sparsity inherent in the problem. Check one of
```

figure windows. nz is a count of the non-zero elements.

'exact' means max(abs(xn-gaussian elimination solution))
'difference' means max(abs(xn-x0)), the difference between two current
iterates.

n	exact	difference
1	5.18e+000	5.00e+000
2	3.75e+000	2.25e+000
3	2.50e+000	1.25e+000
4	1.88e+000	7.50e-001
5	1.25e+000	6.25e-001
6	9.38e-001	3.28e-001
7	6.25e-001	3.13e-001
8	4.69e-001	1.58e-001
9	3.13e-001	1.56e-001
10	2.34e-001	7.84e-002
11	1.56e-001	7.81e-002
12	1.17e-001	3.91e-002
13	7.81e-002	3.91e-002
14	5.86e-002	1.95e-002
15	3.91e-002	1.95e-002
16	2.93e-002	9.77e-003
17	1.95e-002	9.77e-003
18	1.46e-002	4.88e-003
19	9.77e-003	4.88e-003
20	7.32e-003	2.44e-003

****

The matrix of temperatures for the 3D picture contains
the values requested in the problem statement.

```
+-------------------------------->X
|  0        0        0        0   4
|  0   1.42857  2.32143  2.85714  4
|  0   3.39286        5  5.10714  4
|  0   7.14286  9.17857  8.57143  4
|  0        16       16       16  4
Y
```

Temperature values tk requested in the problem are in the table.

t1 = 7.1429
t2 = 9.1786
t3 = 8.5714
t4 = 3.3929
t5 =   5
t6 = 5.1071
t7 = 1.4286
t8 = 2.3214
t9 = 2.8571

Temperatures at the corners are a physical reality, but the averaging principle doesn't use
that information and other considerations dictate how to specify corner temperatures.
That is why you may question the choice of corner temperatures.

EXERCISE SET 5.4
TECHNOLOGY EXERCISES

T1. Use the power method with Euclidean scaling to approximate the dominant eigenvalue and a corresponding eigenvector of A. Choose your own starting vector, and stop when the estimated percentage error in the eigenvalue approximation is less than 0.1 %.

(a) $\begin{bmatrix} 1 & 3 & 3 \\ 3 & 4 & -1 \\ 3 & -1 & 10 \end{bmatrix}$

(b) $\begin{bmatrix} 1 & 0 & 1 & 1 \\ 0 & 2 & -1 & 1 \\ 1 & -1 & 4 & 1 \\ 1 & 1 & 1 & 8 \end{bmatrix}$

The iteration process for part (a) is described in T1a_5_4 and some of the output is shown.

```
>> T1a_5_4
       1   3   3
A =  3   4  -1
       3  -1  10
The relative error for termination is 0.001
The initial approximation vector is
       0.70711
xc =  0.70711
              0
****************
iteration number 1
current relative error 0.1375
    *****        *****
    *****        *****
iteration number 9
current relative error 0.0003056
    *****        *****
The estimated dominant eigenvalue is 10.9083
The corresponding dominant eigenvector is
       0.2931
      -0.0008
       0.9561

Is A*xnew = evalue*xnew?
       3.1589      3.1968
      -0.0802     -0.0090
```

```
    10.4409    10.4294

The eigenvalues of A are
            -1.67063
eig(A) =    5.76102
            10.9096
```

The results for part (b) are similar and details can be reviewed in T1b_5_4.

T2. Repeat Exercise T1, but this time stop when all corresponding entries in two successive eigenvector approximations differ by less than 0.01 in absolute value.

Some details are presented here, and the logic for this approach is provided in the M-file.

```
. >> T2a_5_4
        1   3   3
A = 3   4  -1
        3  -1 10
The absolute error for termination is 0.01
The initial approximation vector is
        0.7071
        0.7071
            0

****************
iteration count 1
current absolute error: 0.24077
        *****         *****
        *****         *****
iteration count 10
current absolute error: 0.006525
        *****         *****
The estimated dominant eigenvalue is 10.9093
A corresponding dominant eigenvector is
                    0.28957
Computed Eigenvector = -0.0073544
                    0.95713
Is the condition Ax = lambda*x close to being satisfied?
        A*x        lambda*x
    3.1389        3.1590
    -0.1178       -0.0802
    10.4474       10.4415

The eigenvalues when eig(A) is used are
            -1.67063
eig(A) =    5.76102
            10.9096
```

As seen in the abbreviated output, the final estimate for the dominant eigenvalue is acceptable. Similar results hold for the other matrix and computations are organized in T2b_5_4.

T3. Repeat Exercise T1 using maximum entry scaling.

Maximum element scaling is a strategy for rescaling an eigenvector estimate that is similar to euclidean scaling.

$$x = \left(1/\max_{k}\left(abs\left(x_k\right)\right)\right)x \text{ - maximum element scaling}$$

$$x = \left(1/\|x\|\right)x \text{ - euclidean scaling}$$

The M-files are almost identical to those in the last problem.

```
>> T3a_5_4
       1   3   3
A =    3   4  -1
       3  -1  10
Power Method With Maximum Entry Scaling
The relative error for termination is 0.001
Compare two most recent eigenvalue estimates as a stopping criterion.
The initial approximation vector is
      1
xc =  1
      0
*****************
iteration number 1
current relative error 0.225
    *****        *****
iteration number 2
current relative error 0.43142
    *****        *****
iteration number 3
current relative error 0.1513
    *****        *****
iteration number 4
current relative error 0.30717
    *****        *****
    *****        *****
iteration number 12
current relative error 0.00055949
    *****        *****
The estimated dominant eigenvalue is 11.889
The corresponding dominant eigenvector is
                   0.29934
computed eigenvector =  -0.013174
                   1
Is A*x close to lambda*x when the current best estimates are used?
```

```
     A*x      lambda*x
   3.2598      3.5589
  -0.1547     -0.1566
  10.9112     11.8890

When eig(A) is used,
            -1.67063
eig(A)  =    5.76102
            10.9096
```

The estimated eigenvalue is quite a bit different than the actual value, and it points out a weakness in the chosen stopping criterion. T3b_5_4 repeats the process for a different matrix.

T4. Suppose that the Google search engine produces 10 internet sites in the search set and that the adjacency matrix for those sites is

$$A = \begin{array}{cc} & \begin{array}{c}\text{Referenced Site}\\ \begin{array}{cccccccccc} 1 & 2 & 3 & 4 & 5 & 6 & 7 & 8 & 9 & 10 \end{array}\end{array} \\ \left[\begin{array}{cccccccccc} 0 & 1 & 1 & 0 & 1 & 1 & 0 & 0 & 0 & 1 \\ 0 & 0 & 1 & 0 & 0 & 0 & 0 & 0 & 0 & 0 \\ 0 & 0 & 0 & 0 & 0 & 0 & 0 & 0 & 0 & 1 \\ 0 & 1 & 1 & 0 & 0 & 1 & 1 & 0 & 0 & 1 \\ 0 & 0 & 0 & 1 & 0 & 0 & 0 & 0 & 0 & 0 \\ 0 & 1 & 0 & 0 & 0 & 0 & 0 & 0 & 0 & 0 \\ 0 & 0 & 0 & 0 & 0 & 0 & 0 & 0 & 1 & 0 \\ 0 & 0 & 0 & 0 & 0 & 1 & 0 & 0 & 0 & 0 \\ 0 & 1 & 1 & 0 & 0 & 1 & 0 & 1 & 0 & 1 \\ 0 & 0 & 0 & 0 & 0 & 1 & 0 & 0 & 0 & 0 \end{array}\right] & \begin{array}{l} 1 \\ 2 \\ 3 \\ 4 \\ 5 \ \text{Referencing Site} \\ 6 \\ 7 \\ 8 \\ 9 \\ 10 \end{array} \end{array}$$

Use the PageRank algorithm to rank the sites in decreasing order of authority for the Google search engine.

The results of the computations are offered because the vectors are too long to display in this manual.

```
>> T4_5_4
```

A table summarizes the authority information after twenty iterations

site	authority
1	0
2	0.47781
3	0.47781
4	0
5	0
6	0.52125
7	0
8	0
9	0
10	0.52125

It appears that sites [1 4 5 7 8 9] can be ignored.
Sites [2 3 6 10] deserve further examination.

There is a way of using the power method to approximate the eigenvalue of A with smallest absolute value when A is an invertible $n \times n$ matrix, and the eigenvalues of A can be ordered according to the size of their absolute values as

$$|\lambda_1| \geq |\lambda_2| \geq \cdots \geq |\lambda_{n-1}| > |\lambda_n|$$

The method uses the fact (proved in Exercise P3 of Section 4.4) that if $\lambda_1, \lambda_2, \ldots, \lambda_n$ are the eigenvalues of A, then $1/\lambda_1, 1/\lambda_2, \ldots, 1/\lambda_n$ are the eigenvalues of A^{-1}, and the eigenvectors of A^{-1} corresponding to $1/\lambda_k$ are the same as the eigenvectors of A corresponding to λ_k. The above inequalities imply that A^{-1} has a dominant eigenvalue of $1/\ell_n$ (why?) which, together with a corresponding eigenvector $\mathbf{x}$ can be approximated by applying the power method to A^{-1}. Once obtained, the reciprocal of this approximation will provide an approximation to the eigenvalue of A that has smallest absolute value, and $\mathbf{x}$ will be a corresponding eigenvector. This technique is called the *inverse power method*. In practice, the inverse power method is rarely implemented by finding A^{-1} and computing successive iterates as

$$\mathbf{x}_k = \frac{A^{-1}\mathbf{x}_{k-1}}{\max(A^{-1}\mathbf{x}_{k-1})} \quad \text{or} \quad \mathbf{x}_k = \frac{A^{-1}\mathbf{x}_{k-1}}{\left\| A^{-1}\mathbf{x}_{k-1} \right\|}$$

Rather, it is usual to let $\mathbf{y}_k = A^{-1}\mathbf{x}_{k-1}$, solve the equation $\mathbf{x}_{k-1} = A\mathbf{y}_k$ for $\mathbf{y}_k$, say by LU-decomposition, and then scale to obtain $\mathbf{x}_k$. The LU-decomposition only needs to be computed once, after which it can be reused to find each new iterate. Use the inverse power method in Exercises T5 and T6.

T5. In Example 6 of Section 4.4, we found the eigenvalues of

$$A = \begin{bmatrix} 3 & 2 \\ 2 & 3 \end{bmatrix}$$

to be $\lambda = 1$ and $\lambda = 5$, and in Example 2 of this section we approximated the eigenvalue $\lambda = 5$ and a corresponding eigenvector using the power method with Euclidean scaling. Use the inverse power method with Euclidean scaling to approximate the eigenvalue $\lambda = 1$ and a corresponding eigenvector. Start with the vector x_0 used in Exercise 2, and stop when the estimated relative error in the eigenvalue is less than 0.001.

Results obtained from T5_5_4 indicate that the inverse power method does not offer a reasonable estimate for the smallest eigenvalue using the parameters and strategy suggested in the text. Changing the error tolerance to 0.0001 doesn't improve matters. An undergraduate research problem is to find out why the algorithm seems to get stuck.

T6. Use the inverse power method with Euclidean scaling to approximate the eigenvalue of

$$A = \begin{bmatrix} 0 & 0 & 2 \\ 0 & 2 & 0 \\ 2 & 0 & 3 \end{bmatrix}$$

The algorithm in T6_5_4 is constructed so that you can experiment with Euclidean scaling or maximum element scaling. The results show that Euclidean scaling gives more reliable results for the matrix in question and the stopping criterion used.

```
>> T6_5_4
        0   0   2
A =     0   2   0
        2   0   3

Inverse power method with Euclidean scaling.
The relative error for termination is 0.001
The initial approximation for an eigenvector vector is
        1
xc =    0
        0
****************
iteration number 4
current relative error 7.1521e-005
    *****        *****
The estimated smallest magnitude eigenvalue is -1
A corresponding computed eigenvector is
        0.8953
xnew =       0
```

```
          -0.44547
Compare both sides of the eigen equation A*x = lambda*x
using the final approximations for x and lambda.
     A*x       lambda*x
  -0.8909     -0.8953
        0           0
   0.4542      0.4455
```

The computed eigenvalues using eig(A) are -1, 2, and 4.

There is a way of using the inverse power method to approximate any eigenvalue λ of a symmetric $n \times n$ matrix A provided one can find a scalar s that is closer to λ than to any other eigenvalue of A. The method is based on the result in Exercise P4 of Section 4.4, which states that if the eigenvalues of A are

$\lambda_1, \lambda_2, ..., \lambda_n$, then the eigenvalues of $A - sI$ are $\lambda_1 - s$, $\lambda_2 - s$, ..., $\lambda_n - s$, and the eigenvectors of $A - sI$ corresponding to $\lambda - s$ are the same as the eigenvectors of A corresponding to λ. Since we are assuming that s is closer to λ than to any other eigenvalue of A, it follows that $1/(\lambda - s)$ is a dominant eigenvalue of the matrix $(A - sI)^{-1}$, so $\lambda - s$ and a corresponding eigenvector $\mathbf{x}$ can be approximated by applying the inverse power method to $A - sI$. Adding s to this approximation will yield an approximation to the eigenvalue λ of A, and $\mathbf{x}$ will be a corresponding eigenvector. This technique is called the ***shifted inverse power*** method.

T7. Given that the matrix

$$A = \begin{bmatrix} 2.0800 & 0.8818 & 0.6235 \\ 0.8818 & 4.0533 & -2.7907 \\ 0.6235 & -2.7907 & 6.0267 \end{bmatrix}$$

has an eigenvalue λ near 3, use the shifted inverse power method with $s = 3$ to approximate λ and a corresponding eigenvector. Choose your own starting vector and stop when the estimated relative error is less than 0.001.

The algorithm in T7_5_4 incorporates a strategy so that you can experiment with Euclidean scaling or maximum element scaling. Convergence seems rapid for the matrix and initial estimates suggested.

```
>> T7_5_4
          2.08     0.8818     0.6235
A =  0.8818     4.0533    -2.7907
     0.6235    -2.7907     6.0267
Shifted inverse power method.
The shift is s = 3
Use Euclidean scaling.
The relative error for termination is 0.001
The initial approximation vector is
       1
xc = 0
       1
****************
iteration number 1
current relative error 0.25949
       *****          *****
iteration number 2
current relative error 0.0010223
       *****          *****
iteration number 3
current relative error 3.6963e-006
       *****          *****
The estimated eigenvalue near s = 3 is 3.159966
A corresponding eigenvector is
          0.70701
xnew = 0.57742
          0.40832
Compare both sides of the eigen equation A*x = lambda*x
using the final estimates for x and lambda.
    A*x         lambda*x
    2.2343       2.2341
    1.8244       1.8246
    1.2903       1.2903

The computed eigenvalues using eig(A) are
1.0000368
 3.159966
7.9999972
```

Command Summary

strvcat   --- concatenate strings vertically

spy   --- generates a profile for a sparse matrix

Chapter 6

Linear Transformations

T1. (a) Find the reflection of the point (1, 3) about the line through the origin of the xy-plane that makes an angle of 12° with the positive x-axis.

The matrix for a reflection is given in the text as

$$T = \begin{bmatrix} \cos(2\theta) & \sin(2\theta) \\ \sin(2\theta) & -\cos(2\theta) \end{bmatrix}$$

and 12° must first be converted to radians. Matrix multiplication Tx using the column vector

$$\mathbf{x} = \begin{bmatrix} 1 \\ 3 \end{bmatrix}$$

produces the desired reflected point. The details are in the M-file T1a_6_1 along with some graphics to indicate the correctness of the approach.

(b) Given the point (2, –1) reflects into the point (–3, 4) about an unknown line L through the origin of the xy-plane, find the reflection of the point $(5, -2)$ about L.

Set $\mathbf{p} = \begin{bmatrix} 2 \\ -1 \end{bmatrix}$, $\mathbf{q} = \begin{bmatrix} -3 \\ 4 \end{bmatrix}$. The goal is to find the reflection matrix T that gets applied to $\mathbf{u} = \begin{bmatrix} 5 \\ -2 \end{bmatrix}$. There is no need to calculate the angle θ if there is a way to calculate the sine and cosine terms in T. We know that $T\mathbf{p} = \mathbf{q}$. Introduce shorthand variables $c2 = \cos(2\theta)$, $s2 = \sin(2\theta)$, and write out Tp symbolically as

135

$$q = Tp = \begin{bmatrix} c2 & s2 \\ s2 & -c2 \end{bmatrix} \begin{bmatrix} p_1 \\ p_2 \end{bmatrix} = \begin{bmatrix} p_1 & p_2 \\ -p_2 & p_1 \end{bmatrix} \begin{bmatrix} c2 \\ s2 \end{bmatrix}$$

This is possible because both T and p are defined by two values. The M-file T1ba_6_1 uses the symbolic toolbox to verify this identity in case you remain unconvinced. The resulting linear system can be solved for c2 and s2 to determine T, and the reflection of u is v = Tu. Details are in T1b_6_1. Visualization is used to reinforce the concepts and to let you see elementary plotting and text insertion techniques.

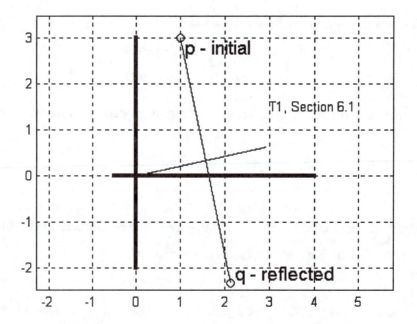

T2. (a) Find the orthogonal projection of the point (1, 3) onto the line through the origin of the xy-plane that makes an angle of 12° with the positive x-axis.

There are several ways to attack this problem, but the text provides a projection matrix P_θ in equation (21) that works nicely. Angles in degrees must be converted to radians before multiplying by (1,3), and the details and graphics are organized in T2a_6_1.

(b) Given the point $(2,-1)$ projects orthogonally onto the point $(-3, 4)$ on an unknown line L through the origin of the xy-plane, find the orthogonal projection of the point $(5,-2)$ on L.

Once the projection matrix P_θ is known, the problem reduces to matrix multiplication.

$$P_\theta = \begin{bmatrix} \cos^2(\theta) & \sin(\theta)\cos(\theta) \\ \sin(\theta)\cos(\theta) & \sin^2(\theta) \end{bmatrix}$$

The point $(-3, 4)$ must lie on the line L, and this information is used to determine the angle needed for the projection. Set

$$q = \begin{bmatrix} -3 \\ 4 \end{bmatrix}$$

Interpret a unit vector along q as defining the angle needed for the projection.

$$(1/\|q\|)q = \begin{bmatrix} \cos(\theta) \\ \sin(\theta) \end{bmatrix} = \frac{1}{5}\begin{bmatrix} -3 \\ 4 \end{bmatrix}$$

Details and some graphics to reinforce the strategy are given in the M-file T2b_6_1.

T3. Generate Figure 17 and explore the behavior of the power sequence for other choices of x_0.

M-file T3_6_1 opens a figure window and displays the behavior of the power sequence for a particular choice of x_0. You can change of x_0 and observe other behaviors by executing the file again. Command window output describes the initial configuration, and the resulting figure window gives an animated view of the iteration process that can't be shown here.

```
>> T3_6_1
A is
    1/2          3/4
   -3/5         11/10
```

```
x0 is
     1
     1
```

EXERCISE SET 6.2
Technology Exercises

T1. Use Formula (13) to find the image of the vector $\mathbf{x} = (1, -2, 5)$ under a rotation of $\theta = \pi/4$ about an axis through the origin oriented in the direction of $\mathbf{u} = \left(\frac{2}{7}, \frac{3}{7}, \frac{6}{7}\right)$.

Formula (13) is direct when a, b, and c are identified.

$$\mathbf{u} = \left(\tfrac{2}{7}, \tfrac{3}{7}, \tfrac{6}{7}\right) = (a, b, c)$$

Rather than enter the nine values for R, you can see patterns in the formula after a while. R is the sum of a symmetric matrix and a skew-symmetric matrix. For instance, the symmetric part is the sum of two parts

$$\left(1-\cos(\theta)\right)\mathbf{u}\mathbf{u}^{T} = \left(1-\cos(\theta)\right)\begin{bmatrix} a \\ b \\ c \end{bmatrix}\begin{bmatrix} a & b & c \end{bmatrix} = \left(1-\cos(\theta)\right)\begin{bmatrix} a^2 & ab & ac \\ ab & b^2 & bc \\ ac & bc & c^2 \end{bmatrix}$$

and

$$\cos(\theta)\begin{bmatrix} 1 & 0 & 0 \\ 0 & 1 & 0 \\ 0 & 0 & 1 \end{bmatrix}$$

The skew-symmetric part is what is left and the details are presented in an M-file.

```
>> T1_6_2
         1
x =    -2
         5

axis vector u is
       2/7
u =    3/7
       6/7
```

```
Rotation Matrix R
       0.73102   -0.57023     0.37477
R =    0.64196    0.7609    -0.094437
      -0.23132    0.30962     0.92229

The rotated vector R*x is
          3.7453
xrot =   -1.352
          3.7609
```

T2. Let

$$A = \begin{bmatrix} -\frac{3}{7} & -\frac{2}{7} & -\frac{6}{7} \\ \frac{6}{7} & -\frac{3}{7} & -\frac{2}{7} \\ -\frac{2}{7} & -\frac{6}{7} & \frac{3}{7} \end{bmatrix}$$

Show that A represents a rotation, and use formulas (16) and (17) to find the axis and angle of rotation.

A quick check is to see if $A^T A = I$ and $\det(A) = 1$. Formulas in (16) and (17) give the cosine of the angle θ and the axis of rotation, respectively. All this information is verified and computed in the M-file T2_6_2. Some of the output is shown below.

```
>> T2_6_2
        -3/7   -2/7   -6/7
A =      6/7   -3/7   -2/7
        -2/7   -6/7    3/7

A*A' is the identity matrix
          1   0   0
A*A' =    0   1   0
          0   0   1

A is an orthognal matrix.
det(A) = 1

A represents a rotation.

cos(theta) = 0.5*(tr(A) - 1) = -0.71429
theta = 2.3664, in radians or
theta = 135.5847, in degrees
```

T3. Let

$$\mathbf{u} = \begin{bmatrix} \frac{1}{3} \\ \frac{2}{3} \\ \frac{2}{3} \end{bmatrix}$$

Use Theorem 6.2.6 to construct the standard matrix for the rotation through an angle of $\pi/6$ about an axis oriented in the direction of $\mathbf{u}$.

Patterns in formula (13) are worth exploiting so as to avoid entering the data one element at a time. Think of the vector u as

$$\mathbf{u} = \begin{bmatrix} \frac{1}{3} \\ \frac{2}{3} \\ \frac{2}{3} \end{bmatrix} = \begin{bmatrix} a \\ b \\ c \end{bmatrix}$$

The matrix in formula (13) is built up from a symmetric part and a skew-symmetric part. The symmetric part is created using an outer product plus a diagonal matrix

$$R_s = (1 - \cos(\theta))\mathbf{u}\mathbf{u}^T + \cos(\theta)\begin{bmatrix} 1 & 0 & 0 \\ 0 & 1 & 0 \\ 0 & 0 & 1 \end{bmatrix}$$

The upper triangular portion of the skew-symmetric part is

$$R_u = \sin(\theta)\begin{bmatrix} 0 & -c & b \\ 0 & 0 & -a \\ 0 & 0 & 0 \end{bmatrix}$$

and the desired standard matrix is $R = R_s + R_u - R_u^T$. Symbolic calculations are offered to convince the skeptics. Note the second arguments in diag when Ru is defined.

```
>> syms a b c t
>> u=[a;b;c];
>> Rs = (1-cos(t))*u*transpose(u) + cos(t)*eye(3);  %symmetric
```

140

```
>> Ru = sin(t)*(diag([-c,-a],1)+diag(b,2)); % upper triangular
>> Rss = Ru - transpose(Ru); % skew-symmetric
>> format compact
>> disp(Rs)
[ (1-cos(t))*a^2+cos(t),          (1-cos(t))*a*b,          (1-cos(t))*a*c]
[          (1-cos(t))*a*b, (1-cos(t))*b^2+cos(t),          (1-cos(t))*b*c]
[          (1-cos(t))*a*c,          (1-cos(t))*b*c, (1-cos(t))*c^2+cos(t)]

>> disp(Ru)
[          0, -sin(t)*c,  sin(t)*b]
[          0,         0, -sin(t)*a]
[          0,         0,         0]

>> disp(Rss)
[          0, -sin(t)*c,  sin(t)*b]
[  sin(t)*c,         0, -sin(t)*a]
[ -sin(t)*b,  sin(t)*a,         0]

>> R = Rs + Rss; % the standard rotation matrix
>> disp(R)

[ (1-cos(t))*a^2+cos(t)   ,(1-cos(t))*a*b-sin(t)*c , (1-cos(t))*a*c+sin(t)*b]
[ (1-cos(t))*a*b+sin(t)*c,    (1-cos(t))*b^2+cos(t), (1-cos(t))*b*c-sin(t)*a]
[ (1-cos(t))*a*c-sin(t)*b, (1-cos(t))*b*c+sin(t)*a,    (1-cos(t))*c^2+cos(t)]
```

This agrees with formula (13) given in the book, and T3_6_2 presents the results of the calculations.

```
>> T3_6_2
A vector u along the rotation axis is
        1/7
u =     2/7
        2/7

The rotation angle in radians is 0.5236
Standard rotation matrix
        0.86876   -0.13739    0.14833
R =     0.14833    0.87696   -0.060492
       -0.13739   0.082365    0.87696
```

EXERCISE SET 6.3
Technology Exercises

T1. Consider the matrix.

$$A = \begin{bmatrix} 2 & 5 & -3 & 7 & 1 & 3 \\ 5 & -2 & 9 & 8 & 4 & -2 \\ -4 & 3 & 8 & 11 & -5 & 2 \\ 11 & 0 & -2 & 4 & 10 & -1 \end{bmatrix}$$

(a) Find the nullspace of A. Express your answer as the span of a set of vectors.

(b) Determine whether the vector $\mathbf{w} = (5, -2, -3, 6)$ is in the range of the linear transformation T_A. If so, find a vector whose image under T_A is $\mathbf{w}$.

A MATLAB command null produces a matrix whose columns are a basis for the nullspace of A, and the M-file T1a_6_3 presents details and a verification for the results of part (a).

```
>> T1a_6_3
          2     5    -3     7     1     3
          5    -2     9     8     4    -2
A =  -4     3     8    11    -5     2
         11     0    -2     4    10    -1

A basis for the null space using nullA = null(A,'r') is
  -264/487        -434/487         47/487
  -863/487         105/487       -302/487
  -478/487          48/487         15/487
      1               0              0
      0               1              0
      0               0              1
A check is to see if A*nullA is zero:
  1.0e-014 *
      0.1776           0              0
          0            0              0
          0            0              0
          0            0        -0.0111
Although not zero, the scaling factor  1.0e-014 *
suggests that roundoff error prevents the computed
matrix from being identically zero. The product is
identically zero when the symbolic toolbox is used.
```

A row reduced form of [A, w] indicates that the system is consistent and that the first three columns of A are a basis for the column space of A. Using this information, a solution of $Ax = w$ are determined with the backslash operator \ and T1b_6_3 presents details that are too extensive to present here.

T2.   Consider the matrix,

$$A = \begin{bmatrix} 3 & -5 & -2 & 2 \\ -4 & 7 & 4 & 4 \\ 4 & -9 & -3 & 7 \\ 2 & -6 & -3 & 2 \end{bmatrix}$$

Show that $T_A : R^4 \to R^4$ is onto in three different ways.

The M-file T2_6_3 presents four different ways using determinants, existence of inverse, row reduced echelon form interpretation, and null space calculations. The output is long and only the first part is given.

```
>> T2_6_3
Show that T(x) = Ax is onto in 4 different ways.
        3 -5 -2   3
       -4  7  4   4
A =     4 -9 -3   7
        2 -6 -3   2
    ***
First way:
det(A) = 45 is not zero
and A is invertible. T is onto.
```

EXERCISE SET 6.4
Technology Exercises

T1. Consider successive rotations of R^3 by 30° about the z-axis, then by 60° about the x-axis, and then by 45° about the y-axis. If it is desired to execute the three rotations by a single rotation about an appropriate axis, what axis and angle should be used?

The strategy is to find the three rotation matrices, carry out the multiplications, and then use formulas (16) and (17) from section 6.2 to determine the angle and axis, respectively. The M-file T1_6_4 is too long to present here, but the output is listed so that you get some idea of the results.

```
>> T1_6_4
Rotation about z-axis
        0.86603      -0.5   0
Rz =    0.5     0.86603   0
```

```
                0        0  1
    ***     ***
Rotation about x-axis
        1          0          0
Rx = 0          0.5    -0.86603
       0    0.86603       0.5
    ***     ***
Rotation about y-axis
        0.70711   0   0.70711
Ry =         0    1         0
        -0.70711  0   0.70711
    ***     ***
Composite rotation matrix R = Ry*Rx*Rz
        0.91856   0.17678    0.35355
R =       0.25    0.43301   -0.86603
        -0.30619  0.88388    0.35355
    ***     ***
    Calculations Related to the Composite Transformation R
The angle in degrees is 69.3559.
The vector x used to determine the axis is
    1
x = 2
    3
A unit vector along the axis of rotation is
        0.93499
v =     0.3525
        0.039124
The unnormalized version of v is v = (R+R')*x + (1 - trace(R))*x
```

T2. (CAS) Consider successive rotations of R^3 through an angle t_1 about the x-axis, then through an angle t_2 about the y-axis, and then through an angle t_3 about the z-axis. Find a single matrix that executes the three rotations.

Required matrices are taken from the text, and abbreviated symbolic results follow. There is nothing unusual about this problem.

```
>> T2_6_4
Rotation about z-axis
[   cos(tz),  -sin(tz),          0]
[   sin(tz),   cos(tz),          0]
[         0,         0,          1]

    ***     ***
Rotation about x-axis
[         1,         0,          0]
[         0,   cos(tx),  -sin(tx)]
[         0,   sin(tx),   cos(tx)]

    ***     ***
```

```
Rotation about y-axis
[  cos(ty),          0,   sin(ty)]
[         0,         1,         0]
[ -sin(ty),          0,   cos(ty)]

   ***    ***
Composite rotation matrix R
```

The output is too long to include here.

```
   ***    ***
```

EXERCISE SET 6.5
Technology Exercises

T1. (a) A figure in the text shows the wireframe of a cube in a rectangular xyz-coordinate system. Find a vertex matrix for the wireframe, and sketch the orthogonal projection of the wireframe on the xy-plane.

(b) Find a 4×4 matrix in homogeneous coordinates that rotates the wireframe 30° about the positive z-axis and then translates the wireframe parallel to the xy-plane by 1 unit in the positive x-direction.

(c) Compute the vertex matrix for the rotated and translated wireframe.

(d) Use the matrix obtained in part (c) to sketch the orthogonal projection on the xy-plane of the rotated and translated wireframe. Does the result agree with your intuitive geometric sense of what the projection should look like?

The results for part (a) are carried out in T1a_6_2, and the resulting projection onto the xy-plane is what you would expect.

```
>> T1a_6_5
Vertex Identification Scheme
1  2  3  4  5  6  7  8
---------------------
0  0  0  0  1  1  1  1
```

145

```
0   1   1   0   0   1   1   0
0   0   1   1   0   0   1   1
*****       *****       *****
The  connectivity  matrix  C  is
1   1   0   1   1   0   0   0
1   1   1   0   0   1   0   0
0   1   1   1   0   0   1   0
1   0   1   1   0   0   0   1
1   0   0   0   1   1   0   1
0   1   0   0   1   1   1   0
0   0   1   0   0   1   1   1
0   0   0   1   1   0   1   1
C  is  symmetric  because  C  -  C'  is  0

projected  vertices  onto  xy-plane
        0       0       0       0       1       1       1       1
        0       1       1       0       0       1       1       0
        0       0       0       0       0       0       0       0
```

The matrix with homogenous coordinates that carries out the composite transformation is evaluated and described in T1b_6_5.

```
>> T1b_6_5
Translation matrix:
        1   0   0   1
        0   1   0   0
T = 0   0   1   0
        0   0   0   1

Rotation matrix:
        0.86603         -0.5   0   0
            0.5   0.86603   0   0
R =         0             0   1   0
            0             0   0   1

The  matrix  for  rotation  followed  by  translation  is
        0.86603         -0.5   0   1
            0.5   0.86603   0   0
T*R =       0             0   1   0
            0             0   0   1
```

Results of third part include graphics and T1c_6_5 provides details.

```
>> T1c_6_5
The  vertex  matrix  for  the  rotated  and  translated  wireframe  is

1.00   0.50   0.50   1.00   1.87   1.37   1.37   1.87
0.00   0.87   0.87   0.00   0.50   1.37   1.37   0.50
0.00   0.00   1.00   1.00   0.00   0.00   1.00   1.00

Check  out  the  wireframe  in  the  figure  window.
Use  the  rotation  tool  in  the  toolbar  of  the  figure  window
```

to manipulate the figure even further.

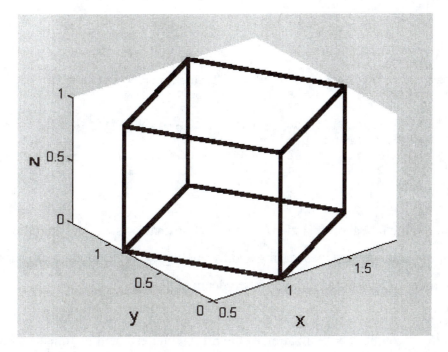

T2. Repeat parts (b), (c), and (d) of Exercise T1 assuming that the wireframe is rotated 30° about the positive *y*-axis and then translated so the vertex at the origin moves to the point (1, 1, 1).

Rotating about the positive y-axis introduces a different rotation matrix that is part of the output of the M-file T2_6_5. Graphical output includes two figures that show the object rotated and translated, and then projected onto the xy-plane.

```
>> T2_6_5
Translation matrix:
        1   0   0   1
        0   1   0   1
T =  0   0   1   1
        0   0   0   1

Rotation matrix:
      0.86603   0         0.5   0
            0   1           0   0
R =     -0.5   0   0.86603   0
```

```
            0  0          0  1
```

The matrix for rotation followed by translation is

```
         0.86603  0        0.5  1
               0  1          0  1
T*R =       -0.5  0    0.86603  1
               0  0          0  1
```

The vertex matrix for the rotated and translated wireframe is

```
               1  1    1.5    1.5  1.866  1.866  2.366  2.366
vertices = 1   2    2      1      1      2      2      1
           1   1    1.866  1.866  0.5    0.5    1.366  1.366
```

The vertex matrix for the projection is

```
            1  1  1.5  1.5  1.866  1.866  2.366  2.366
Wproj = 1   2  2    1    1      2      2      1
        0   0  0    0    0      0      0      0
```

T3. A figure in the text shows a letter *L* in an *xy*-coordinate system and an italicized version of that letter created by shearing and translating. Use the method of Example 3 to find the vertices of the shifted italic *L* to two decimal places.

M-file T3_6_5 provides a visualization of the original letter L and its transformed version. Coordinates for the letter are taken from that sketch, and coordinates of the sheared letter are part of the output. Shearing requires a parameter that is given by the tangent of the angle shown in the sketch. These details are in the M-file.

```
>> T3_6_5
Each column of the next table gives vertex
coordinates of the original letter L

x| 0.00  5.00  5.00  1.50  1.50  0.00
y| 0.00  0.00  1.50  1.50  7.00  7.00

Each column of the next table gives vertex
coordinates of the sheared and translated L

x| 1.50  6.50  6.76  3.26  4.23  2.73
y| 0.00  0.00  1.50  1.50  7.00  7.00
```

The y coordinates of the vertices of the transformed letter are the same as the originals because a horizontal shear and translation was applied.

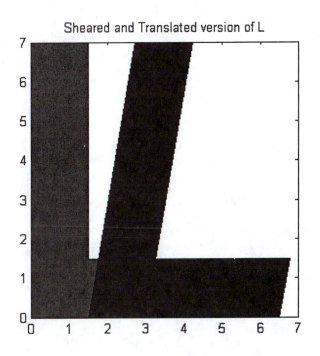

Sheared and Translated version of L

Command Summary

Wireframe3   --- function to assist in 3D presentation of wireframe

Chapter 7

Dimension and Structure

EXERCISE SET 7.1
TECHNOLOGY EXERCISES

T1. Are any of the vectors in the set

$$S = \{(2, 6, 3, 4, 2), \ (3, 1\ 5, 8, 3), \ (5, 1, 2, 6, 7), \ (8, 4, 3, 2, 6), \ (5, 5, 6, 3, 4)\}$$

linear combinations of predecessors? Justify your answer.

One strategy is to form a matrix V of the column vectors v_k mentioned above and decide whether the system $Vx = 0$ has nontrivial solutions. If so, then at least one column is a linear combination of previous ones. Otherwise, the columns are linearly independent.

```
>> T1_7_1
                         2   3   5   8   5
                         6   1   1   4   5
V = [v1,v2,v3,v4,v5]  =  3   5   2   3   6
                         4   8   6   2   3
                         2   3   7   6   4

The row-reduced form R is
      1   0   0   0   0
      0   1   0   0   0
R  =  0   0   1   0   0
      0   0   0   1   0
      0   0   0   0   1
```

R has 5 pivots and so the columns of V are
linearly independent. No column of V can be a
linear combination of any other columns.

T2. (CAS) Find the exact canonical basis (no decimal approximations) for the solution space of $Ax = 0$, given that

$$A = \begin{bmatrix} 1 & 5 & 2 & 4 & 4 & 7 \\ 3 & 2 & 4 & 9 & 1 & 3 \\ 5 & 2 & 4 & 8 & 5 & 7 \\ 9 & 9 & 10 & 21 & 10 & 17 \end{bmatrix}$$

This can be accomplished in two ways because the entries are small integers. One way uses the operator null for a numeric matrix and the other is to use a symbolic matrix. A third way is to use rref and interpret the results, but that is not done here. See T2_7_1.

```
>> T2_7_1
        1    5    2    4    4    7
        3    2    4    9    1    3
A =  5    2    4    8    5    7
        9    9   10   21   10   17
```

A basis for the null space nul(A) obtained from nullA = null(A,'r') consists of the columns below.

```
              1/2      -2       -2
             3/16    -9/8    -13/8
           -87/32   29/16    25/16
nullA =        1       0        0
               0       1        0
               0       0        1
        *****        *****
```

A times this matrix should be the zero matrix.

```
              0   0   0
              0   0   0
A*nullA =  0   0   0
              0   0   0
        *****        *****
```

The symbolic toolbox can be used.
A basis for the null space obtained from null(sym(A)) consists of the columns below.

```
            [     1,    8/3,    7/3]
            [     0,      1,      0]
            [-29/2,  -29/2,    -22]
nullsym = [     6,   16/3,   26/3]
            [     1,      0,      0]
            [     0,      0,      1]
        *****        *****
```

A times this matrix should be the zero matrix.

```
              [0, 0, 0]
              [0, 0, 0]
A*nullsym = [0, 0, 0]
              [0, 0, 0]
        *****        *****
```

The two presentations are correct but different because solutions of a homogeneous system are not unique.

EXERCISE SET 7.2
TECHNOLOGY EXERCISES

T1. Devise a procedure for using your technology utility to find the dimension of the subspace spanned by a set of vectors in R^n, and use all of your procedures to find the dimension of the subspace of R^5 spanned by the vectors

$$\mathbf{v}_1 = (2, 2, -1, 0, 1).\ \mathbf{v}_2 = (-1, -1, 2. -3, 1),\ \mathbf{v}_3 = (1, 1, -2, 0, -1),\ \mathbf{v}_4 = (0, 0\ 1, 1, 1)$$

A method is described in T1_7_2.

```
>> T1_7_2
                          2 -1   1   0
                          2 -1   1   0
V = [v1,v2,v3,v4]  =  -1   2  -2   1
                          0 -3   0   1
                          1   1  -1   1
*** Method 1 ***
The row reduced form R is
        1   0   0    1/3
        0   1   0   -1/3
R =     0   0   1    -1
        0   0   0     0
        0   0   0     0

There are 3 pivot columns in the row reduced form
and this is the dimension of the subspace spanned by the columns.
```

T2. Let $S = \{\mathbf{v}_1, \mathbf{v}_2, \mathbf{v}_3, \mathbf{v}_4, \mathbf{v}_5\}$, where

$$\mathbf{v}_1 = (1, 2, 1),\quad \mathbf{v}_2 = (4, 4, 4),\quad \mathbf{v}_3 = (1, 0, 1),\quad \mathbf{v}_4 = (2, 4, 2),\quad \mathbf{v}_5 = (0, 1, 1)$$

Find all possible subsets of S that are bases for R^3.

The strategy selected for this problem is to form a matrix from the five column vectors and then calculate the dimension N of the column space of that matrix. Based on this count, a search can be conducted to find which groups of N column vectors are linearly independent. No search is necessary if the dimension is less than three. The code in T2_7_2 is written so that other sets of vectors or matrices can be entered and analyzed.

```
>> T2_7_2
Find all subsets of the columns of A that
constitute a basis for R3.
*****************************************
```

```
        1   4   1   2   0
A  =  2   4   0   4   1
        1   4   1   2   1
```

Any basis for the column space of A has 3 vectors
because rref(A) has 3 pivot columns 1 2   5.

Any basis for the column space of A is also
a basis for the subspace spanned by the original
vectors.
***   ***   ***   ***
The rows of the next matrix give the columns
of A to be examined for linear independence.

```
                                    1   2   3
                                    1   2   4
                                    1   2   5
                                    1   3   4
                                    1   3   5
Combinations of columns to consider = 1   4   5
                                    2   3   4
                                    2   3   5
                                    2   4   5
                                    3   4   5
```

```
        *****        *****
A(:,1   2   5) is a basis for the column space of A and hence R3.
***    ***    ***    ***
A(:,1   3   5) is a basis for the column space of A and hence R3.
***    ***    ***    ***
A(:,2   3   5) is a basis for the column space of A and hence R3.
***    ***    ***    ***
A(:,2   4   5) is a basis for the column space of A and hence R3.
***    ***    ***    ***
A(:,3   4   5) is a basis for the column space of A and hence R3.
***    ***    ***    ***
```

T3. (CAS) Use a determinant test to find conditions on a, b, c, and d under which the
vectors

$$\mathbf{v}_1 = (a, b, c, d), \quad \mathbf{v}_2 = (-b, a, d, -c), \quad \mathbf{v}_3 = (-c, -d, a, b), \quad \mathbf{v}_4 = (-d, c, -b, a)$$

form a basis for R^4.

A symbolic determinant handles this problem with no difficulty.

```
>> T3_7_2
The vectors are the rows of A.
        [ a,    b,    c,    d]
        [-b,    a,    d,   -c]
A =   [-c,   -d,    a,    b]
        [-d,    c,   -b,    a]
```

154

The vectors v1, v2, v3, and v4 are linearly independent because
det(A) = (a^2+b^2+d^2+c^2)^2 is not zero unless a, b, c, and d are
all zero. They constitute a basis because they are linearly
independent vectors in R4.

EXERCISE SET 7.3
TECHNOLOGY EXERCISES

T1. Many technology utilities provide a command for finding a basis for the nullspace of
a matrix.

(a) Determine whether your utility has this capability, and, if so use that command to
find a basis for the null space of the matrix

$$A = \begin{bmatrix} 3 & 2 & 1 & 3 & 5 \\ 6 & 4 & 3 & 5 & 7 \\ 9 & 6 & 5 & 7 & 9 \\ 3 & 2 & 0 & 4 & 8 \end{bmatrix}$$

(b) Confirm that the basis obtained in part (a) is consistent with the basis that results
when your utility is used to find the general solution of the linear system $A\mathbf{x} = \mathbf{0}$.

The command null(A) or null(A,'r') produces a matrix whose columns are a
basis for the null space of A, and the second version requests that the output be
determined by first row reducing A and interpreting the results. It should be what you get
when working out the problem with pencil and paper. The first version returns unit
vectors that are mutually orthogonal. The output is an empty array when the null space is
the zero subspace, and no second argument is used when A is a symbolic matrix.

```
>> T1_7_3
      3   2   1   3   5
      6   4   3   5   7
A = 9   6   5   7   9
      3   2   0   4   8
A has small entries and the command null(A,'r')
returns a matrix whose columns are a basis
for the null space of A.

A basis for the null space of A is found in the columns of
          -2/3   -4/3   -8/3
           1      0      0
nullA =    0      1      3
```

155

```
                0    1    0
                0    0    1
Premultiplication by A should give a zero matrix.
             0  0  0
             0  0  0
A*nullA =  0   0  0
             0  0  0
```

T2. Some technology utilities provide a command for finding a basis for the row space of a matrix.

(a) Determine whether your utility has this capability, and, if so use that command to find a basis for the row space of the matrix

$$A = \begin{bmatrix} 2 & -1 & 3 & 5 \\ 4 & -3 & 1 & 3 \\ 3 & -2 & 3 & 4 \\ 4 & -1 & 15 & 17 \\ 7 & -6 & -7 & 0 \end{bmatrix}$$

(b) Confirm that the basis obtained in part (a) is consistent with the basis that results when your utility is used to find the reduced row echelon form of A.

MATLAB provides no immediate command for determining a basis for the row space of a matrix. The non-zero rows of the row reduced form of A are a basis for the row space of A. So are the corresponding rows of A.

```
>> T2_7_3
        2   -1    3    5
        4   -3    1    3
A = 3   -2    3    4
        4   -1   15   17
        7   -6   -7    0

The row reduced form is
        1   0   0   6
        0   1   0   7
R = 0   0   1   0
        0   0   0   0
        0   0   0   0
The non-zero rows of R are 1  2  3.

Linear combinations of A(1  2  3,:) reproduce R(1  2  3,:)
using a weight matrix xr that links the two via
xr*A(1  2  3,:) = R(1  2  3,:).
```

```
The weight matrix is
        3.5    1.5   -4
xr =    4.5    1.5   -5
       -0.5   -0.5    1
and
                      1   0   0   6
xr*A(1   2   3,:) =   0   1   0   7
                      0   0   1   0

This agrees with the non-zero rows of R
                   1   0   0   6
R(1   2   3,:) =   0   1   0   7
                   0   0   1   0
and it means that rows 1   2   3 of A or of R
are linearly independent. They also span the
row space of A as the following calculations reveal.
```

Additional calculations show that these rows or A or R also span the row space. The M-file gives all the details.

T3. Some technology utilities provide a command for finding a basis for the column space of a matrix.

(a) Determine whether your utility has this capability, and, if so use that command to find a basis for the column space of the matrix in Exercise T2.

(b) Confirm that the basis obtained is consistent with the basis that results when your utility is used to find a basis for the row space of A^T.

One strategy is to row reduce A and select the columns of A that correspond to the pivot columns of the row reduced form. A second way is to use the command orth(A) which calculates a basis for the column space of A. The output of orth consists of orthogonal unit vectors, and the underlying strategy is quite unlike the first one. The help file for orth provides additional details. Another approach is to calculate and suitably interpret a singular value decomposition of A, and this topic is discussed in Chapter 8. The M-file T3_7_3 lays out the details and no output is provided.

T4. Determine which of the vectors $\mathbf{b}_1 = (6, -15, -14, -16)$, $\mathbf{b}_2 = (1, 3, 17, -9)$, and $\mathbf{b}_3 = (-10, 9, -14, -48)$ if any, lie in the subspace of R^5 spanned by the vectors in Example 4.

One strategy is to form a matrix A whose columns are the vectors in Example 4 and check for consistency in the three systems $Ax_k = b_k$. Consistency is decided by verifying that no pivot occurs in the last column of the row reduced form of the augmented matrix $[A, b_k]$. A partial listing of the output of T4_7_3 indicates the nature of the computations.

```
>> T4_7_3
        1  -2    0    2
        0   1    5   10
A =  0  -3  -14  -28
        0  -2   -9  -18
        2  -4    0    4
                        2    1    7
                        6   16   14
B = [b1,b2,b3] =  -17  -45    2
                       -11  -29    1
                         4    2   14
Which columns of B are in the column space of A?
The row reduced form R1 of [A,b1] is
        1   0   0   2   4
        0   1   0   0   1
R1 = 0   0   1   2   1
        0   0   0   0   0
        0   0   0   0   0

b1 is in the column space of A
The weights x such that A*x1 = b1 are
        4
        1
x1 = 1
        0
```

EXERCISE SET 7.4
TECHNOLOGY EXERCISES

T1. These are questions dealing with the rank of a matrix.

(a) Some technology utilities provide a command for finding the rank of a matrix. Determine whether your utility has this capability, and if so, use that command to find the rank of the matrix in Example 1.

158

(b) Confirm that the rank obtained in part (a) is consistent with the rank obtained by using your utility to find the number of nonzero rows in the reduced row echelon form of the matrix.

The rank function rank(A) calculates the number of linearly independent rows or columns of A. It returns a count and does not indicate which columns constitute a basis for the column space. Although simple in principle, it is a delicate matter to calculate the rank of a matrix in a reliable fashion. Strategies used in calculating rank(A) make it reliable within the limits of floating point arithmetic.

```
>> T1_7_4
          1     0     0     0     2
         -2     1    -3    -2    -1
A =       0     5   -14    -9     0
          2    10   -28   -18     4
The calculated rank of A using rank(A) is 3


                  1     0     0     0     2
                  0     1     0     1   -42
R = rref(A) =  0     0     1     1   -15
                  0     0     0     0     0


There are 3 pivot rows in R.
This number is consistent with that found using rank(A).
```

T2. Most technology utilities do not provide a direct command for finding the nullity of a matrix since the nullity can be computed using the rank command and Formula (2). Use that method to find the nullity of the matrix in Exercise T1 of Section 7.3, and confirm that the result obtained is consistent with the number of basis vectors obtained in that exercise.

Routine calculations are provided in T2_7_4.

```
>> T2_7_4
          1     0     0     0     2
         -2     1    -3    -2    -1
A =       0     5   -14    -9     0
          2    10   -28   -18     4

The computed rank of A using rank(A) is 3
and A has 5 columns.
The nullity of A is computed as 5 - rank(A) = 2
```

159

```
This can be verified by using the command null(A) and counting
the number of columns in the output.
size(null(A),2) = 2
```

T3. Confirm Formula (2) for some 5×7 matrices of your choice.

Formula (2) is $\text{rank}(A) + \text{nullity}(A) = n$, and the M-file T3_7_4 verifies it for a matrix. No output is offered.

T4. **Sylvester's rank inequalities** (whose proofs are somewhat detailed) state that if A is a matrix with n columns and B is a matrix with n rows, then

$$\text{rank}(A) + \text{rank}(B) - n \leq \text{rank}(AB) \leq \text{rank}(A)$$

$$\text{rank}(A) + \text{rank}(B) - n \leq \text{rank}(AB) \leq \text{rank}(B)$$

Confirm these inequalities for some matrices of your choice.

Only one matrix pair is verified in T4_7_4. Output shown below uses rank several times.

```
>> T4_7_4
        1   2   3   4   5
A =     2   3   4   5   6

        1   2
        2   3
B =     3   4
        4   5
        5   6

Does rank(A) + rank(B) - n <= rank(AB) <= rank(A)?
rank(A) = 2
rank(B) = 2
rank(A*B) = 2
   ***     ***     ***
2 + 2 - 5 <= 2 <= 2
n = 5 in this case.
   *****       *****
Does rank(A) + rank(B) - n <= rank(AB) <= rank(B)?
2 + 2 - 5 <= 2 <= 2
```

T5.

(a) Consider the matrices

$$A = \begin{bmatrix} 7 & 4 & -2 & 4 \\ 2 & -3 & 7 & -6 \\ 5 & 6 & 2 & -5 \\ 3 & 3 & -5 & 8 \end{bmatrix} \quad \text{and} \quad B = \begin{bmatrix} 7.1 & 4 & -2 & 4 \\ 2 & -3 & 7 & -6 \\ 5 & 6 & 2 & -5 \\ 3 & 3 & -5 & 8 \end{bmatrix}$$

that differ in only one entry. Compute A^{-1} and use the result in Exercise P6 to compute B^{-1}.

(b) Check your result in computing B^{-1} directly.

Translating the mathematics into code is the hardest part of this problem, and T5_7_4 is heavily commented to assist you.

```
>> T5_7_4
        7   4  -2   4
        2  -3   7  -6
A =  5   6  -2   5
        3   3  -5   8
A plus the next matrix, an outer product, is B
        0.1   0   0   0
          0   0   0   0
u*v'  =   0   0   0   0
          0   0   0   0
      *****        *****
The inverse of B using inv(B) is
            0.25255    0.013292   -0.15064   -0.022153
           -0.095259  -0.083961    0.28489   -0.1934
inv(B)  =  -0.35224    0.27093     0.26274    0.21511
           -0.27913    0.19584     0.11387    0.34027

% build the inverse of B using the formula
Ainv = inv(A);
Binv = Ainv - (1/(1 + v'*Ainv*u))*(Ainv*u)*(v'*Ainv);
echo off
Using the formula, the computed inverse of B is
            0.25255    0.013292   -0.15064   -0.022153
           -0.095259  -0.083961    0.28489   -0.1934
Binv  =    -0.35224    0.27093     0.26274    0.21511
           -0.27913    0.19584     0.11387    0.34027
```

EXERCISE SET 7.5
TECHNOLOGY EXERCISES

T1. (***Finding rank using determinants***) Since a square matrix is invertible if and only its determinant is nonzero, it follows from Exercise P5 that the rank of a nonzero matrix A is the order of the largest square submatrix of A whose determinant is nonzero. Use this fact to find the rank of A, and check your answer using a different method to find the rank.

$$A = \begin{bmatrix} 3 & -1 & 3 & 2 & 5 \\ 5 & -3 & 2 & 3 & 4 \\ 1 & -3 & -5 & 0 & -7 \\ 7 & -5 & 1 & 4 & 1 \end{bmatrix}$$

Starting with the largest square matrices to be found in A, a search is started for the first sub-matrix with a non-zero determinant. This is a laborious process because all combinations of rows and columns must be considered. A function nchoosek provides a convenient way to select all the combinations of rows and columns needed in the search process. For example, all combinations of the integers 1, 2, 3, 4 taken three at a time are listed by the rows of

```
>> comb = nchoosek(1:4,3)
comb =

    1    2    3
    1    2    4
    1    3    4
    2    3    4
```

Suitable calls to nchoosek are an essential part of the search process in T1_7_5.

```
>> T1_7_5
% Find the rank of a matrix by looking for a largest
% submatrix with a non-zero determinant.
echo off
        3 -1   3   2   5
        5 -3   2   3   4
A =     1 -3  -5   0  -7
        7 -5   1   4   1
    ***     ***     ***
*** The determinant of every square submatrix of size 4 is zero. ***
The rank of A is 3
Select rows 1   2   4 and columns 1   2   5
for a largest sub-matrix with non-zero determinant.
The submatrix is
```

```
                  3 -1   5
A(1  2  4,1  2  5) = 5 -3   4
                  7 -5   1
```
Its determinant is 8.
The calculated rank of A using rank(A) is 3

EXERCISE SET 7.6
TECHNOLOGY EXERCISES

T1. Consider the vectors

$$\mathbf{v}_1 = (1, 2, 4, -6, 11, 23, -14, 0, 2, 2) \quad \mathbf{v}_2 = (3, 1, -1, 7, 9, 13, -12, 8, 6, -30)$$

$$\mathbf{v}_3 = (5, 5, 7, -5, 31, 59, -40, -8, 10, 1), \mathbf{v}_4 = (5, 0, -6, 20, 7, 3, -47, 16, 10, -8)$$

Use Algorithm 1 to find a subset of these vectors that forms a basis for

span$\{\mathbf{v}_1, \mathbf{v}_2, \mathbf{v}_3, \mathbf{v}_4\}$ and express those vectors not in the basis as linear combinations of

basis vectors.

Algorithm 1 suggested forming a matrix A associated with the above vectors and then
computing the row reduced form. The pivot columns of the row reduced form act as
pointers to the corresponding columns of A. This is what is done in T1_7_6, and use is
made of a special feature of `rref` that returns the pivot columns as part of the output. It
turns out that the first three columns are a basis and $\mathbf{v}_4 = -\mathbf{v}_1 + 2\mathbf{v}_2$.

T2. Consider the matrix

$$A = \begin{bmatrix} 1 & 3 & 2 & 1 \\ -2 & -6 & 0 & -6 \\ 3 & 9 & 1 & 8 \\ -1 & -3 & -3 & -6 \\ 1 & 3 & 2 & 1 \\ 4 & 12 & 1 & 11 \end{bmatrix}$$

(a) Use Algorithm 1 to find a subset of the column vectors of A that forms a basis for the column space of A, and express each column vector of A that is not in that basis as a linear combination of the basis vectors.

(b) By applying Algorithm 1 to A^T, find a subset of the row vectors of A that forms a basis for the row space of A, and express each row vector that is not in the basis as a linear combination of the basis vectors.

(c) Use Algorithm 2 to find a basis for the nullspace of A^T.

This starts out looking like the previous problem, but expressing each column vector of A that is not in that basis requires that we also know the pivot rows. Pivot rows tell which row elements of the row reduced form are to be used as weights for verifying the spanning property. Part (a) is taken up in T2a_7_6, and so forth for the other two parts. Output is too long to put here, but the code is written so that other matrices can be substituted for the requested one. A small example is shown here.

```
> T2a_7_6
        1   2   3   4
A = 0   0   1   5

The row reduced form R is
        1   2   0 -11
R = 0   0   1   5

the pivot columns are 1   3,
and the pivot rows are 1   2.

All are important in verifying the spanning property of a basis.
   ***     ***     ***     ***     ***
Columns 1   3 of A are a basis for col(A),
and the actual columns are
                1   3
A(:,[1   3]) = 0   1
     *****         *****
The remaining columns of A are linear combinations
of these vectors.

                              2
A(:,2)  = A(:,[1   3])*R([1   2],2)  = 0

                              4
A(:,4)  = A(:,[1   3])*R([1   2],4)  = 5
```

164

The remaining parts in T2b_7_6 and T2c_7_6 are variations on a theme introduced in the first part.

EXERCISE SET 7.7
TECHNOLOGY EXERCISES

T1. (*Standard matrix for an orthogonal projection*). Most technology utilities do not have a special command for computing orthogonal projections, so several commands must be used in combination. One way to find the standard matrix for the orthogonal projection onto a subspace W spanned by a set of vectors $\{v_1, v_2, \ldots, v_k\}$ is to first find a basis for W, then create a matrix A that has the basis vectors as columns, and then use Formula (27).

(a) Find the standard matrix for the orthogonal projection of R^4 onto the subspace W spanned by

$$v_1 = (1, 2, 3, -4), \quad v_2 = (2, 3, -4, 1), \quad v_3 = (2, -5, 8, -3), \quad v_4 = (5, 26, -9, -12),$$
$$v_5 = (3, -4, 1, 2).$$

(b) Use the matrix obtained in part (a) to find the $\text{proj}_W x$, where $x = (1, 0, -3, 7)$.

(c) Find $\text{proj}_{W^\perp} x$ for the vector in part (b).

```
>> T1_7_7
        1    2    2    5    3
        2    3   -5   26   -4
A =     3   -4    8   -9    1
       -4    1   -3  -12    2

The vector x to project onto W is
        1
        0
x =    -3
        7

   ***    ***
Use the pivot columns from [R,pivotcols] = rref(A) to obtain
M = A(:,pivotcols), a matrix whose columns are a basis for col(A).
pivotcols = 1   2   3

        1    2    2
        2    3   -5
M =     3   -4    8
       -4    1   -3
```

```
Matrix for orthogonal projection onto W = col(A)

                         0.99915   -0.014407   -0.016102   -0.019492
                        -0.014407    0.75508    -0.27373    -0.33136
P = M*inv(M'*M)*M' =    -0.016102   -0.27373     0.69407    -0.37034
                        -0.019492   -0.33136    -0.37034     0.55169

As a check, P*P should be P.

            0.99915   -0.014407   -0.016102   -0.019492
           -0.014407    0.75508    -0.27373    -0.33136
P'*P =     -0.016102   -0.27373     0.69407    -0.37034
           -0.019492   -0.33136    -0.37034     0.55169


    ***     ***
Orthogonal projection of x onto W
           0.91102
          -1.5127
xproj =   -4.6907
           4.9534

Orthogonal projection of x onto the perp of W
           0.088983
           1.5127
xperp =    1.6907
           2.0466

As a check, the dot product of the projections should be zero.
dot(xproj,xperp) = 0
```

T2. Confirm that the following linear system is consistent, and find the solution that lies in the row space of the coefficient matrix:

$$
\begin{aligned}
12x_1 + 14x_2 - 15x_3 + 23x_4 + 27x_5 &= 5 \\
16x_1 + 18x_2 - 22x_3 + 29x_4 + 37x_5 &= 8 \\
18x_1 + 20x_2 - 21x_3 + 32x_4 + 41x_5 &= 9 \\
10x_1 + 12x_2 - 16x_3 + 20x_4 + 23x_5 &= 4
\end{aligned}
$$

A solution of $AA^T c = b$ has the property that $x = A^T c$ not only solves the system, but $x^T = c^T A$ is in the row space of A. Details are in the M-file T2_7_7.

T3. (CAS) Use Formula (16) to compute the standard matrix for the orthogonal projection of R^3 onto the line spanned by the nonzero vector $\mathbf{a} = (a_1, a_2, a_3)$, and then confirm that the resulting matrix has the properties stated in Theorem 7.7.6.

The symbolic output is too extensive to display here, but the computations in T3_7_7 follow the discussion in Section 7.7.

EXERCISE SET 7.8
TECHNOLOGY EXERCISES

T1. (*Least Squares Fit to Data*) Most technology utilities provide a direct command for fitting various kinds of functions to data by minimizing the sum of the squares of the deviations between the function values and the data. Use this command to check the result in Example 7.

Previously, we used $\mathtt{x} = \mathtt{A \backslash b}$ to solve a system with a square coefficient matrix. The functionality of this same backslash operator includes solving least squares problems when the A has full column rank. The help file tells more.

```
>> T1_7_8
% The back slash operator \ carries out least
% squares too. It isn't necessary to form
% the normal equations.
x1 = M\y; % use backslash
x2 = (M'*M)\(M'*y); % use normal equations
% x1 and x2 are the same.
echo off
        1   0
        1   1
M = 1   2
        1   3


        1
        3
y = 4
        4


            1.5
x1 = M\y =   1
```

```
x2 = (M'*M)\(M'*y) =      1
```

T2. (*Least Squares Solutions of Linear Systems*) Some technology utilities provide a direct command for finding least squares solutions of linear systems, whereas others require that you set up and solve the associated normal system. Determine how your utility handles this problem and check the result in Example 3.

The backslash operator \ is discussed in T1 and this problem gives another verification of the basic ideas. No output is offered, but check T2_7_8.

T3. Find the least-squares fit in Exercise 13 by solving the normal system, and then compare the result to that obtained by using the direct command for a least-squares fit.

Once the design matrix for the cubic fit is created, both solutions are immediate. This is almost the same as T1 and T2. Although T3_7_8 addresses the problem, the M-file T3a_7_8 uses the function `polyfit` instead of the normal equations. Graphics appear in both cases.

T4. *Pathfinder* is an experimental, lightweight, remotely-piloted, solar-powered aircraft that was used in a series of experiments by NASA to determine the feasibility of applying solar power for long-duration, high-altitude flight. In August of 1997 Pathfinder recorded the following data relating altitude H and temperature T. Show, by plotting the data, that a linear model is reasonable and then find the least-squares line $H = T_0 + kT$ of best fit.

Altitude (thousands of feet)	15	20	25	30	35	40	45
Temperature (Celsius)	4.5	−5.9	−16.1	−27.6	−39.8	−50.2	−62.9

A linear model for altitude leads to a simple least squares problem with details that appear in T4_7_8. The data plot and fitted line are shown in the figure window generated in the M-file.

Three important models in applications are:

Exponential models $(y = ae^{bx})$

Power function models $(y = ax^b)$

Logarithmic models $(y = a + b \ln x)$

where a and b are to be determined to fit experimental data as closely as possible. Exercises T5 – T7 are concerned with a procedure, called *linearization*, by which the data is transformed to a form in which a straight-line fit by least squares can be used to approximate the constants.

T5. Transforming the data before making a least squares fit is a powerful tool in data analysis.

(a) Show that making the substitution $Y = \ln y$ in the equation $y = ae^{bx}$ produces the equation $Y = bx + \ln a$ whose graph in the xY-plane is a line of slope b and Y-intercept $\ln a$.

(b) Part (a) suggests that a curve of the form $y = ae^{bx}$ can be fitted to n data points (x_i, y_i) by letting $X_i = \ln x_i$, and $Y_i = \ln y_i$, then fitting a straight line to the transformed data points (X_i, Y_i) by least squares to find b and $\ln a$, and then computing a from $\ln a$. Use this method to fit a power-function model to the following data, and graph the curve and data in the same coordinate system.

x	0	1	2	3	4	5	6	7
y	3.9	5.3	7.2	9.6	12	17	23	31

T6. Other transformations can be used in least squares fitting.

(a) Show that making the substitutions $X = \ln x$ and $Y = \ln y$ and in the equation

$y = ax^b$ produces the equation $Y = bX + \ln a$ whose graph in the XY-plane is a line of slope b and Y-intercept $\ln a$.

(b) Part (a) suggests that a curve of the form $y = ae^{bx}$ can be fitted to n data points (x_j, y_j) by letting $X_i = \ln x_i$, and $Y_i = \ln y_i$, then fitting a straight line to the transformed data points (X_i, Y_i) by least squares to find b and $\ln a$, and then computing a from $\ln a$. Use this method to fit a power-function model to the following data, and graph the curve and data in the same coordinate system.

x	2	3	4	5	6	7	8	9
y	1.75	1.91	2.03	2.13	2.22	2.30	2.37	2.43

The effort in this problem is to set up the appropriate least squares model after transforming the data. Plotting is easy with MATLAB, and T6_7_8 provides graphics and many details as to how it is done.

T7. Logarithmic substitutions sometimes help understand the underlying data.

(a) Show that making the substitution $X = \ln x$ in the equation $y = a + b \ln x$ produces the equation $Y = a + bX$ whose graph in the Xy-plane is a line of slope b and y-intercept a.

(b) Part (a) suggests that a curve of the form $y = a + b \ln x$ can be fitted to n data points (x_i, y_i) by letting $X_i = \ln x_i$ and then fitting a straight line to the transformed data points (X_i, y_i) by least squares to find b and a. Use this method to fit logarithmic model to the data, and graph the curve and data in the same coordinate system.

x	2	3	4	5	6	7	8	9
y	4.07	5.30	6.21	6.79	7.32	7.91	8.23	8.51

Except for a slightly different data transformation in x, this problem is identical to the previous one. T7_7_8 provides a running commentary on the strategy and only graphics appear.

T8. (*Center of a circle by least squares*) Least squares methods can be used to estimate the center (h, k) of a circle $(x - h)^2 + (y - k)^2 = r^2$ using measured data points on its circumference. To see how, suppose that the data points are

$$(x_1, y_1), \ (x_2, y_2), \ \ldots, \ (x_n, y_n)$$

and rewrite the equation of the circle in the form

$$(*) \quad 2xh + 2yk + s = x^2 + y^2$$

where

$$(**) \quad s = r^2 - h^2 - k^2$$

Substituting the data points in (*) yields a linear system in the unknowns h, k, and s, which can be solved by least squares to estimate their values. Equation (**) can then be used estimate r. Use this method to approximate the center and radius of a circle from the following measured data points on the circumference:

x	19.880	20.919	21.735	23.375	24.361	25.375	25.979
y	68.874	67.676	66.692	64.385	62.908	61.292	60.277

[*Note*: The data in this problem is based on archaeological excavations of a circular starting line for race track in the Greek city of Corinth dating from about 500 B.C. For a more detailed discussion of the problem and its history, see the article *Finding the Center of a Circular Starting Line in an Ancient Greek Stadium*, SIAM Review, Vol. 39, No. 4 by Chris Rorres and David Gilman Romano.]

Graphics appear when this M-file is executed.

```
>> T8_7_8
% Set up the design matrix for a least squares fit
% Model 2xh + 2yk + s = x^2+y^2;    r^2 = s + h^2 + k^2
X = [2*x, 2*y, ones(n,1)];
b = x.^2 + y.^2;
hks = X\b; % least squares solution
%
h = hks(1); k = hks(2); % coordinates of center
s = hks(3);
r = (s + h^2 + k^2)^.5; % radius
echo off
       x         y
    19.8800    68.8740
    20.9190    67.6760
    21.7350    66.6920
    23.3750    64.3850
    24.3610    62.9080
    25.3750    61.2920
    25.9790    60.2770

Center at (-18.3534,35.4513)
Radius is 50.7952
```

EXERCISE SET 7.9
TECHNOLOGY EXERCISES

T1, (*Gram-Schmidt Process*) Most technology utilities provide a command for performing some variation of the Gram-Schmidt process to produce either an orthogonal or orthonormal set. Some utilities require the starting vectors to be linearly independent, while others allow the set to be linearly dependent. In the latter case the utility eliminates linearly dependent vectors and produces an orthogonal or orthonormal basis for the space spanned by the original set. Determine how your utility performs the Gram-Schmidt process and use it check the results obtained in Example 7.

An orthogonal decomposition $A = QR$ is equivalent to the Gram-Schmidt process: Q is an orthogonal matrix, R is upper triangular, and the columns of Q are the vectors that arise during the Gram-Schmidt process. Suppose A has two columns and use partitioned matrices for $A = QR$

$$A = [a_1, a_2] = [q_1, q_2]\begin{bmatrix} r_{11} & r_{12} \\ 0 & r_{22} \end{bmatrix} = [r_{11}q_1, r_{12}q_1 + r_{22}q_2]$$

Comparing columns, we see that $a_1 = r_{11}q_1$ and $a_2 = r_{12}q_1 + r_{22}q_2$, or equivalently

$$r_{11} = \|a_1\|, \ q_1 = (1/r_{11})a_1, \ r_{22}q_2 = a_2 - r_{12}q_1$$

A little more thought shows that $r_{12} = q_1 \bullet a_2$ and r_{22} is a normalizing constant that ensures that q_2 is a unit vector. When written out, it looks exactly like the Gram-Schmidt process outlined in the text. This process can be carried out using a MATLAB function qr. There are two versions, and the one that reproduces the Gram-Schmidt process is [Q,R] = qr(A,0). The second argument of zero causes Q to have just as many columns as A, and the upper triangular matrix is adjusted accordingly. Omitting the second argument produces an orthogonal matrix with as many rows as A.

```
>> T1_7_9
          16/25     0      -12/25
A =           0   1/25        0
         -12/25     0       9/25

[Q,R] = qr(A,0); produces the economy size data

       -0.8   0   0.6
Q =       0   1     0
        0.6   0   0.8

       -0.8      0   0.6
R =       0   0.04     0
          0      0     0

           0.64      0   -0.48
A = Q*R =     0   0.04       0
          -0.48      0    0.36
```

The final products for A look a little different
than A because of formatting.

Up to a scaling factor, the first column of Q agrees with the first column of A. That scaling factor is found in the (1,1) element of R. Column 3 is in the span of the first two and this is detected by noticing that the (3,3) element of R is zero. The third column of R shows that $a_3 = 0.8*q_1$. Another M-file qrtest provides a platform for reviewing the different versions of qr.

T2, (*Normalization*) Some technology utilities have a command for normalizing a set of nonzero vectors. Determine whether your utility has such a command, and, if so, use it to convert the following set of orthogonal vectors to an orthonormal set:

$$\mathbf{v}_1 = (2, 1, 3, -1), \quad \mathbf{v}_2 = (3, 2, -3, -1), \quad \mathbf{v}_3 = (1, 5, 1, 10)$$

There is no immediate MATLAB command for normalizing a collection of nonzero vectors. There is a way to accomplish this using the powerful primitives in MATLAB. The M-file T2_7_9 provides details.

```
>> T2_7_9
                       2   3   1
                       1   2   5
V= [v1,v2,v3]   =      3  -3   1
                      -1  -1  10

The columns of Vn are normalized versions of the
corresponding columns of V.
                              0.5164     0.62554    0.088736
                              0.2582     0.41703    0.44368
Vn = V*diag(1./(sum(V.^2)).^.5)  =   0.7746    -0.62554    0.088736
                             -0.2582    -0.20851    0.88736

As a check, take the norm of v1 and multiply it by Vn(:,1)
and see if it agrees with V(:,1).
                                          0
                                          0
V(:,1)  -  norm(V(:,1))*Vn(:,1)  =  0
                                          0
```

T3. Find an orthonormal basis for the subspace of R^7 spanned by the vectors
$$\mathbf{w}_1 = (1, 2, 3, 4, 5, 6, 7), \quad \mathbf{w}_2 = (1, 0, 3, 1, 1, 2, -2), \quad \mathbf{w}_3 = (1, 4, 3, 7, 9, 10, 1)$$

The function `orth` is designed to give just such a basis once the vectors are collected as the columns of a matrix. No details of T3_7_9 are provided here.

T4. Find orthonormal bases for the four fundamental spaces of the matrix

$$A = \begin{bmatrix} 2 & -1 & 3 & 5 \\ 4 & -3 & 1 & 3 \\ 3 & -2 & 3 & 4 \\ 4 & -1 & 15 & 17 \\ 7 & -6 & -7 & 0 \end{bmatrix}$$

Some of the subspaces can collapse to the zero subspace, and this complicates the general problem of reporting results. MATLAB function `orth` and `null` are used to generate decimal representations of orthonormal bases for the four subspaces. Only part of the output of T4_7_9 is shown here.

```
>> T4_7_9
        2    -1     3     5
        4    -3     1     1
A =     3    -2     3     4
        4    -1    15    17
        7    -6    -7     0

Orthonormal basis for the column space of A:
              -0.24338     -0.12631     0.24472      -0.35171
              -0.090184    -0.34737    -0.84346      -0.39972
orth(A)  =    -0.22291      -0.2171    -0.28648       0.84348
              -0.93119    -0.0034276    0.13473      -0.071221
               0.1258      -0.90346     0.35842     0.00044013

Orthonormal basis for the null space of the transpose of A:
              -0.86095
                     0
null(A') =    -0.33113
               0.33113
               0.19868

They should be orthogonal complements and the next product should be a
zero vector.
nulAt'*colA = 0   0   0   0
    ***      ***       ***       ***
```

T5. **(CAS)** Find the standard matrix for orthogonal projection onto the subspace of R^4 spanned by the nonzero vector $\mathbf{w} = (a, b, c, d)$. Confirm that the matrix is symmetric and idempotent, as guaranteed by Theorem 7.7.6, and use Theorem 7.9.3 to confirm that it has rank 1.

Set up symbolic variables and use the symbolic toolbox in order to understand the nature of this problem. Outer products play a role.

```
>> T5_9_9
      [x1]         [a]
      [x2]         [b]
x =   [x3]    w =  [c]
      [x4]         [d]

Projection of x onto w
             [(a*x1+b*x2+c*x3+d*x4)/(a^2+b^2+c^2+d^2)*a]
             [(a*x1+b*x2+c*x3+d*x4)/(a^2+b^2+c^2+d^2)*b]
projw_x =    [(a*x1+b*x2+c*x3+d*x4)/(a^2+b^2+c^2+d^2)*c]
             [(a*x1+b*x2+c*x3+d*x4)/(a^2+b^2+c^2+d^2)*d]
   ***     ***     ***     ***
Projection matrix associated with projection of x onto w.
Remove a scaling factor for display purposes.
                               [a^2, a*b, a*c, a*d]
                               [a*b, b^2, b*c, b*d]
(a^2+b^2+c^2+d^2)*proj_mat =   [a*c, b*c, c^2, c*d]
                               [a*d, b*d, c*d, d^2]
By inspection, proj_mat is symmetric. It has rank 1 because
it is essentially an outer product. Outer products almost always
have rank 1.

proj_mat is idempotent:
                             [0, 0, 0, 0]
                             [0, 0, 0, 0]
proj_mat^2 - proj_mat =      [0, 0, 0, 0]
                             [0, 0, 0, 0]
```

EXERCISE SET 7.10
TECHNOLOGY EXERCISES

T1. Most linear algebra technology utilities have a command for finding QR-decompositions. Use that command to find the QR-decompositions of the matrices in Examples 1 and 2.

A *QR*-decomposition is discussed in connection with the Gram-Schmidt process in section 7.9, and the M-file T1_7_10 gives details for this particular problem.

T2. Construct a *QR*-decomposition of the matrix

$$A = \begin{bmatrix} 1 & 1 & 1 \\ 1 & 0 & 2 \\ 0 & 1 & 2 \end{bmatrix}$$

by applying the Gram-Schmidt process to the column vectors to find Q and using Formula (5) to compute R. Compare your result to that produced by the command for computing *QR*-decompositions.

The only difference is a consistent minus sign that is due to the algorithm used in qr. M-file T2_7_10 has a symbolic switch so that you can watch a symbolic version of Gram-Schmidt being executed.

```
>> T2_7_10
        1   1   1
A  =    1   0   2
        0   1   2

First unit vector
        0.70711
q1 =    0.70711
            0
Second unit vector
        0.40825
q2 =   -0.40825
        0.8165
Third unit vector
       -0.57735
q3 =    0.57735
        0.57735
    ***     ***     ***
                    0.70711     0.40825    -0.57735
Q = [q1,q2,q3]  =   0.70711    -0.40825     0.57735
                        0        0.8165     0.57735

Orthogonality check:
Q'Q is
            1   0   0
Q'Q =       0   1   0
            0   0   1
```

```
                             -0.70711     0.40825   -0.57735
Q from qr(A) =  -0.70711    -0.40825    0.57735
                            0    0.8165    0.57735
     ***     ***     ***


The upper triangular R is
             1.4142   0.70711   2.1213
R = Q\A =       0    1.2247   1.2247
                0       0    1.7321
and
             -1.4142   -0.70711   -2.1213
R from qr(A) =      0    1.2247   1.2247
                    0       0    1.7321
```

T3. Consider the linear system

$$x_1 + 5x_2 + 3x_3 = 0.8$$
$$x_1 + 3x_2 + 4x_3 = 0.8$$
$$x_1 + x_2 + 5x_3 = 0.6$$
$$x_1 + 2x_2 + x_3 = 0.4$$

Find the least squares solution of the system using the command provided for that purpose. Compare your result to that obtained by finding a *QR*-decomposition of the coefficient matrix and applying the method of Example 2.

An abbreviated version of T3_7_10 shows that the results are the same.

```
>> T3_7_10
        1   5   3
        1   3   4
A = 1   1   5
        1   2   1


          4/5
          4/5
b =      3/5
          2/5
     ***     ***     ***
Least squares solutions of Ax = b obtained in two ways:
               0.1381
xls = A\b = 0.090476
               0.080952


                  0.1381
xqr = R\(Q'*b) = 0.090476
                  0.080952
```

178

T4. Use Householder reflections and the method of Example 7 to find a *QR*-decomposition of the matrix

$$A = \begin{bmatrix} 1 & 2 & 1 \\ -1 & -2 & 3 \\ 0 & 4 & 5 \end{bmatrix}$$

Compare your answer to that produced by the command provided by your utility for finding *QR*-decompositions.

Outlined in the text is a conceptual development of an algorithm for computing a Householder reflection. When translated into MATLAB, the result is the function `housemat`.

```
function Q = housemat(x)
% Q = housemat(x)
% Purpose: Generate a Householder reflection based on
% a non-zero column vector x

%Think of x as being v in the book and
% w as ||x||*[1; zeros(n-1,1)

x = x(:); % make sure it is a unit vector
n = length(x); % number of elements
x(1) = x(1) - norm(x); % This is a = v - w in the text

% Q = I - (2/a'*a)*a*a' is next
t = 2/dot(x,x);
Q = eye(n) - t*(x*x'); % Use the formula in the text
```

A vector from Example 5 helps make the point.

```
>> x = [1 2 2]';
>> Q = housemat(x)
Q =
      1/3              2/3              2/3
      2/3              1/3             -2/3
      2/3             -2/3              1/3

>> disp(formatA(Q'*Q,'Q''*Q')) % Q is orthogonal
          1   0   0
Q'*Q =    0   1   0
          0   0   1
>> disp(formatA(Q*x,'Q*x'))
          3
Q*x =     0
          0
```

Things are not quite what they seem. With a different format scheme, the elements of Qx that are reported to be zero are not because of roundoff errors.

```
>> format short e
>> y = Q*x
y =

   3.0000e+000
   2.2204e-016
   1.1102e-016
```

This is why it may not be a bad to idea to add the code

```
>> y(2:end) = 0
y =
    -3
     0
     0
```

and remove the fluff due to roundoff error if you use housemat and are really expecting those elements to be zero. M-file T4_7_10 takes you through the QR-decomposition using Householder reflections and offers a comparison with the output of qr.

T5. (CAS) Show that if $\mathbf{a} = (a, b, c)$ is a nonzero vector in R^3, then the standard matrix H for the reflection of R^3 about the hyperplane $\mathbf{a}^\perp$ is

$$H = \begin{bmatrix} 1 - \dfrac{2a^2}{a^2+b^2+c^2} & -\dfrac{2ab}{a^2+b^2+c^2} & -\dfrac{2ac}{a^2+b^2+c^2} \\ -\dfrac{2ab}{a^2+b^2+c^2} & 1 - \dfrac{2b^2}{a^2+b^2+c^2} & -\dfrac{2bc}{a^2+b^2+c^2} \\ -\dfrac{2ac}{a^2+b^2+c^2} & -\dfrac{2bc}{a^2+b^2+c^2} & 1 - \dfrac{2c^2}{a^2+b^2+c^2} \end{bmatrix}$$

This looks almost obvious when done with paper and pencil: $H = I - \dfrac{2}{\mathbf{a} \bullet \mathbf{a}} \mathbf{a}\mathbf{a}^T$. Doing it with the Symbolic Toolbox has some surprises that are documented in T5_7_10. The moral of the story is that simplification rules give correct results that can look nothing like the hand calculations.

EXERCISE SET 7.11
TECHNOLOGY EXERCISES

T1. Transition matrices provide a consistent way to express vectors in different bases.

(a) Confirm that $B_1 = \{u_1, u_2, u_3, u_4, u_5\}$ and $B_2 = \{v_1, v_2, v_3, v_4, v_5\}$ are bases for R^5 and find the transition matrices $P_{B_1 \to B_2}$ and $P_{B_2 \to B_1}$.

$$u_1 = (3, 1, 3, 2, 6), \quad u_2 = (4, 5, 7, 2, 4), \quad u_3 = (3, 2, 1, 5, 4),$$
$$u_4 = (2, 9, 1, 4, 4), \quad u_5 = (3, 3, 6, 6, 7)$$

$$v_1 = (2, 6, 3, 4, 2), \quad v_2 = (3, 1, 5, 8, 3), \quad v_3 = (5, 1, 2, 6, 7),$$
$$v_4 = (8, 4, 3, 2, 6), \quad v_5 = (5, 5, 6, 3, 4)$$

(b) Find the coordinate matrices with respect to B_1 and B_2 of $w = (1, 1, 1, 1, 1)$.

Notation is the hardest part of this problem. Once a vector in list form is written as a column vector, it becomes the coordinate matrix with respect to the standard basis E that consists of the columns of I_5, the identity matrix of size five. In fact,

$$[u_1]_E = \begin{bmatrix} 3 \\ 1 \\ 3 \\ 2 \\ 6 \end{bmatrix} \text{ and } [u_1]_B = \begin{bmatrix} 1 \\ 0 \\ 0 \\ 0 \\ 0 \end{bmatrix}$$

and the matrix U is a transition matrix $P_{B_1 \to E}$.

$$U = \left[[u_1]_E, [u_2]_E, [u_3]_E, [u_4]_E, [u_5]_E \right]$$

The same can be said for the other vectors

$$P_{B_2 \to E} = V = \left[[v_1]_E, [v_2]_E, [v_3]_E, [v_4]_E, [v_5]_E \right]$$

Move to matrix equations as quickly as possible. This choice leads to the relations

$$[x]_E = U[x]_{B_1} \text{ and } [x]_E = V[x]_{B_2} \text{ when } x = (x_1, x_2, x_3, x_4, x_5)$$

The connection between the bases comes from an interpretation of the resulting identity

$$V[x]_{B_2} = U[x]_{B_1}$$

The result $[x]_{B_2} = V^{-1}U[x]_{B_1}$ means that $V^{-1}U = P_{B_1 \rightarrow B_2}$ and subsequently $U^{-1}V = P_{B_2 \rightarrow B_1}$.

A quick check verifies the first choice. Just watch the subscripts.

$$V^{-1}U[\mathbf{u}_1]_{B_1} = V^{-1}U \begin{bmatrix} 1 \\ 0 \\ 0 \\ 0 \\ 0 \end{bmatrix} = V^{-1}[\mathbf{u}_1]_E = V^{-1}V[\mathbf{u}_1]_{B_2} = [\mathbf{u}_1]_{B_2}$$

The output of T1_7_11 should make more sense now. Everybody is happy when a discussion of bases transition matrices is finished. Those working in computer graphics finally get comfortable with the details.

T2. An important problem in many applications is to perform a succession of rotations to align the positive axes of a right-handed xyz-coordinate with the corresponding axes of a right-handed XYZ-coordinate system that has the same origin (see Figure Ex T3 in the text). This can be accomplished by three successive rotations, which, if needed, can be composed into a single rotation about an appropriate axis. The three rotations involve angles θ, φ, and ψ, called **Euler angles**, and a vector $\mathbf{n}$, called the **axis of nodes**. As indicated in the figure, the axis of nodes is orthogonal to both the z-axis and Z-axis and hence is a along the line of intersection of the xy- and XY-planes, θ is the angle from the positive z-axis to the positive Z-axis, φ is the angle from the positive x-axis to the axis of nodes, and ψ is the angle from the axis of nodes to the positive X-axis. The positive xyz-axes can be aligned with the positive XYZ-axes by first rotating the xyz-axes counterclockwise about the positive z-axis through the angle φ to carry the positive x-axis into the axis of nodes, then rotating the resulting axes counterclockwise about the axis of nodes through the angle θ to carry the positive z-axis into the positive Z-axis, and then rotating the resulting axes counterclockwise through the angle ψ about the positive Z-axis to carry the axis of nodes into the positive X-axis. Suppose that a rectangular xyz-coordinate system and a rectangular XYZ-coordinate system have the same origin and that $\theta = \pi/6$, $\varphi = \pi/3$, and $\psi = \pi/4$ are Euler angles. Find a matrix A that relates the xyz-coordinates and XYZ-coordinates of a fixed point related by

$$\begin{bmatrix} X \\ Y \\ Z \end{bmatrix} = A \begin{bmatrix} x \\ y \\ z \end{bmatrix}$$

The sketch in the text has to be analyzed carefully. If $w = (x, y, z)$ is a typical vector in 3-

space, then $[w] = \begin{bmatrix} x \\ y \\ z \end{bmatrix}$ is the coordinate matrix of w with respect to standard basis. We are

first told to rotate counterclockwise about the z-axis by ϕ radians, and the appropriate rotation matrix is

$$R_\phi = \begin{bmatrix} \cos(\phi) & -\sin(\phi) & 0 \\ \sin(\phi) & \cos(\phi) & 0 \\ 0 & 0 & 1 \end{bmatrix}$$

From the picture in the text, the coordinates of the axis of nodes is

$$\mathbf{n} = R_\phi e_1 = \begin{bmatrix} \cos(\phi) & -\sin(\phi) & 0 \\ \sin(\phi) & \cos(\phi) & 0 \\ 0 & 0 & 1 \end{bmatrix} \begin{bmatrix} 1 \\ 0 \\ 0 \end{bmatrix}$$

The columns of R_ϕ give the coordinates of the rotated x-axis otherwise known as the axis of nodes $\mathbf{n}$ for this problem, the rotated y-axis, and the rotated z-axis, respectively, using the original, standard basis. In fact, $[w] = R_\phi [w]_\phi$, and the subscripted vector $[w]_\phi$ gives the coordinates of w in the rotated system. This is a subtle point, and you should make sure you understand the last sentence. In the once-rotated system, we want to rotate counterclockwise by θ about its x-axis (axis of nodes) and a rotation matrix for such is

183

$$R_\theta = \begin{bmatrix} 1 & 0 & 0 \\ 0 & \cos(\theta) & -\sin(\theta) \\ 0 & \sin(\theta) & \cos(\theta) \end{bmatrix}$$

But what do we apply it to, and how is it interpreted? Using the same reasoning as before, $[w]_\phi = R_\theta [w]_\theta$. The last coordinate change involves a counterclockwise rotation by ψ about the z-axis in the twice-rotated coordinate system. The appropriate matrix is

$$R_\psi = \begin{bmatrix} \cos(\psi) & -\sin(\psi) & 0 \\ \sin(\psi) & \cos(\psi) & 0 \\ 0 & 0 & 1 \end{bmatrix}$$

and it remains to understand how it gets applied. The insight gained after the first rotation tells us that $[w]_\theta = R_\psi [w]_\psi$ should hold. The problem stated in the text is asking for

$[w]_\psi = A[w]$ or for $[w] = A[w]_\psi$, depending on how you read it. We take the former, and it is a matter of putting it all together as $[w] = R_\phi [w]_\phi = R_\phi R_\theta [w]_\theta = R_\phi R_\theta R_\psi [w]_\psi$.

Recalling that the three rotation matrices are orthogonal, we arrive at $[w]_\psi = R_\psi^T R_\theta^T R_\phi^T [w]$

and finally $A = R_\psi^T R_\theta^T R_\phi^T$. This makes perfect sense in retrospect. Substitute for the angles to get a numerical version of A.

```
          -0.39952     0.80801    0.43301
A =       -0.80801    -0.53349       0.25
           0.43301       -0.25    0.86603
```

T2plot_7_11 carries out the numerical calculations needed for this problem and gives a 3D sequence of snapshots so that you can follow the rotations. An associated M-file T2sym_7_11 generates a symbolic version that agrees with A.

Respective columns of the related matrix $R_\phi R_\theta R_\psi$ give the coordinate matrices of the standard unit vectors in the final XYZ coordinate system in terms of the standard basis we started out with. Think about it. The reason for the extended discussion is to show

how coordinate matrices can be used to reason about a geometric problem. Consult a good book on computer graphics for additional insight because there are several ways to approach the problem.

Command Summary

qr   --- *QR*-decompositions of a matrix

rank   --- calculate the rank of a matrix using a singular value decomposition

null   --- calculate the null space of a matrix

orth   --- calculate an orthonormal basis for the column space of a matrix

housemat   --- apply a Householder reflection to a column vector

nchoosek   --- generate combinations of objects based on input arguments

Chapter 8

Diagonalization

T1. Let $T: R^5 \rightarrow R^3$ be the linear operator given by the formula

$$T(x_1, x_2, x_3, x_4, x_5) = (7x_1 + 12x_2 - 5x_3, \; 3x_1 + 10x_2 + 13x_4 + x_5, \; -9x_1 - x_3 - 3x_5)$$

and let $B = \{\mathbf{v}_1, \mathbf{v}_2, \mathbf{v}_3, \mathbf{v}_4, \mathbf{v}_5\}$ and $B' = \{\mathbf{v}_1{}^z, \mathbf{v}_2{}^z, \mathbf{v}_3{}^z\}$ be the bases for R^5 and R^3 in

which

$$\mathbf{v}_1 = (1, 1, 0, 0, 0), \; \mathbf{v}_2 = (0, 1, 1, 0, 0), \; \mathbf{v}_3 = (0, 0, 1, 1, 0), \; \mathbf{v}_4 = (0, 0, 0, 1, 1),$$
$$\mathbf{v}_5 = (1, 0, 0, 0, 1)$$

and

$$\mathbf{v}_1{}^z = (1, 2, -1), \; \mathbf{v}_2{}^z = (2, 1, 3), \; \mathbf{v}_3{}^z = (1, 1, 1)$$

(a) Find the matrix $[T]_{B',B}$.

(b) For the vector $\mathbf{x} = (3, 7, -4, 5, 1)$ find compute $[\mathbf{x}]_B$ and use the matrix obtained in part (a) to compute $[T(\mathbf{x})]_{B'}$.

It is always a challenge to keep subscripts straight when working through a problem of this type. The vectors are given in list form, and when they are written as column vectors they become coordinate vectors relative to a standard basis. For example, when E_3 is the standard ordered basis for R^3

$$\begin{bmatrix} 1 \\ 2 \\ -1 \end{bmatrix} = \left[\mathbf{v}_1{}^z \right]_{E_3} \qquad E_3 = \left\{ \begin{bmatrix} 1 \\ 0 \\ 0 \end{bmatrix}, \begin{bmatrix} 0 \\ 1 \\ 0 \end{bmatrix}, \begin{bmatrix} 0 \\ 0 \\ 1 \end{bmatrix} \right\}$$

This provides a way of keeping track of the details that are given in the M-file T1_8_1.

T2. Let $[T]$ be the standard matrix for the linear transformation in Exercise T1 and let B and B' be the bases in that exercise. Find the factorization in Formula (28).

Matrix $[T]$ is called A in the M-file T1_8_1. Also, the transition matrix from B to the standard basis is called B, and the transition matrix from B' to the standard ordered basis for R^3 is called Bp. The matrix identity $\mathrm{Bp}[T]_{B',B} = \mathrm{AB}$ shows how the two matrices for T are related. It is apparent that $V = \mathrm{Bp}$ and $U = \mathrm{B}$ when these results are compared with Formula (28).

EXERCISE SET 8.2
TECHNOLOGY EXERCISES

T1. Most linear algebra technology utilities have specific commands for diagonalizing a matrix. If your utility has this capability, then you may find it described as a "Jordan decomposition" or by some similar name involving the work "Jordan". Use this command to diagonalize the matrix in Example 4.

The matrix in question is

$$A = \begin{bmatrix} 0 & 0 & -2 \\ 1 & 2 & 1 \\ 1 & 0 & 3 \end{bmatrix}$$

and `eig` is the main tool for diagonalizing a matrix. The command $[P,D] = eig(A)$ produces a matrix P and a diagonal matrix D that satisfies $AP = PD$. The diagonal elements of D are eigenvalues of A, and the corresponding columns of P are eigenvectors. There is no assurance that the columns of P are linearly independent which is necessary when discussing diagonalization. You can check their linear independence by checking if the number of rows agrees with `rank(P)`. A Jordan decomposition is notoriously difficult to compute because of its delicate nature. Even if you could compute it in principle, there is no way to deal with general matrices of size five or greater. No MATLAB command is available for calculating the Jordan decomposition. Nevertheless, the matrix A can be diagonalized.

```
>> T1_8_2
For the command [P,D] = eig(A), the diagonal
elements of D are eigenvalues of A and the
columns of P are corresponding eigenvectors.
The columns of P are normalized to have unit length.

ev = eig(A), returns the eigenvalues in ev
   ***    ***    ***    ***


      0   0  -2
A = 1   2   1
      1   0   3


      0  -0.8165    0.70711
P = 1   0.40825          0
      0   0.40825   -0.70711


      2   0   0
D = 0   1   0
      0   0   2


rank(P) = 3

P*D*P^-1 should agree with A
                0   0  -2
A = P*D*P^-1 = 1   2   1
                1   0   3
It does.
```

T2. (a) Show that the matrix

$$A = \begin{bmatrix} -13 & -60 & -60 \\ 10 & 42 & 40 \\ -5 & -20 & -18 \end{bmatrix}$$

is diagonalizable by finding the nullity of $\lambda I - A$ for each eigenvalue λ and calling on an appropriate theorem.

Even with a small matrix having integer entries, calculating eigenvectors using the definition can be interesting.

```
>> T2_8_2
A is
    -13    -60    -60
     10     42     40
     -5    -20    -18
```

The computed eigenvalues in the array evals are
     2.00000000000007
     6.99999999999991
     2.00000000000000

Use the command null(evals(k)*eye(3) - A) to get null space basis
Basis for null space corresponding to evals(1)
     0.81658314566575
    -0.49726538754411
     0.29311960112767

Basis for null space corresponding to evals(2)
     0.80178372573727
    -0.53452248382485
     0.26726124191243

Basis for null space corresponding to evals(3)
     0.98473192783466   0
    -0.12309149097933   -0.70710678118655
    -0.12309149097933   0.70710678118655

This is odd because we have four eigenvectors for a 3x3 matrix.
It turns out that these four vectors are linearly dependent.
The first vector is a linear combination of the two vectors
in the third group even though the computed eigenvalvues
appear to be distinct.
This shows how subtle it can be to work with floating point arithmetic.

Use [P,D] = eig(A) and get eigenvectors in columns of P
D is
     2.00000000000007   0   0
                    0   6.99999999999991   0
                    0   0   2.00000000000000

P is
    -0.87287156094397   -0.80178372573727   -0.94287315745881
     0.43643578047199   0.53452248382485   -0.08611224175843
    -0.21821789023599   -0.26726124191242   0.32183053112313

rank(P) = 3
The rank result shows that the columns of P are
linearly independent and constitute a basis for R3.

A different analysis shows that the exact eigenvalues are 2, 2, 7

(b) Find a basis for R^3 consisting of eigenvectors of A.

A computed basis is found in the columns of P as described previously, but more can be
said. The characteristic polynomial of A has integer entries because A has only integer
entries. The round command in the next code fragment removes the effect of roundoff

errors. The subsequent plot shows that 2 is a double root. Once the exact roots are known, the command `null` can be used with precise data to obtain another basis.

```
>> p = round(poly(A)); % poly(A) returns accurate coefficients
>> disp(p) % polynomial, MATLAB style
     1    -11    32    -28
>> disp(polyval(p,[2,7])) % This shows that 2, 7 are exact roots.
     0     0
>> t = linspace(1,7.5); % 100 data points for graph
>> plot(t,polyval(p,t)) % plot characteristic polynomial
>> hold on;plot(2,0,'o',7,0,'o') % mark roots
>> title('Characteristic Polynomial')
>> hold off
% Use exact eigenvalues to get clean eigenvectors
>> disp(null(A - 2*eye(3),'r'))
    -4    -4
     1     0
     0     1

>> disp(null(A - 7*eye(3),'r'))
     3
    -2
     1
```

These three vectors are a basis too.

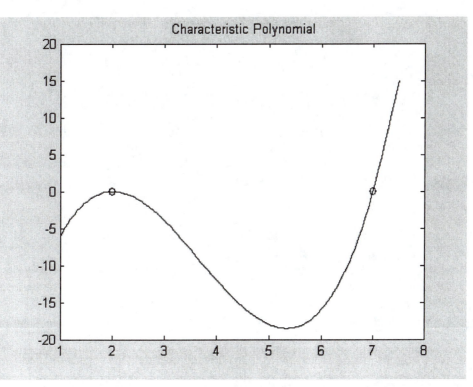

191

T3. Construct a 4 × 4 diagonalizable matrix A whose entries are nonzero and whose characteristic equation is $p(\lambda) = (\lambda - 2)^2(\lambda + 3)^2$, and check your result by diagonalizing A.

Set $A = PDP^{-1}$ where P is the inverse of a Hilbert matrix.

$$P = \begin{bmatrix} 16 & -120 & 240 & -140 \\ -120 & 1200 & -2700 & 1680 \\ 240 & -2700 & 6480 & -4200 \\ -140 & 1680 & -4200 & 2800 \end{bmatrix} \quad D = \begin{bmatrix} -3 & 0 & 0 & 0 \\ 0 & -3 & 0 & 0 \\ 0 & 0 & 2 & 0 \\ 0 & 0 & 0 & 2 \end{bmatrix}$$

Consult `help invhilb` for details. Matrix A has all the desired properties as the M-file output demonstrates.

```
>> T3_8_2
         16  -120    240   -140
       -120  1200  -2700   1680
P =     240 -2700   6480  -4200
       -140  1680  -4200   2800

       -3   0   0   0
        0  -3   0   0
D =     0   0   2   0
        0   0   0   2

                    222      160    123.33333       100
                  -2400    -1698        -1300     -1050
A = P*D*P^-1 =     5550     3900         2977      2400
                  -3500    -2450    -1866.6667     -1503

Computed eigenvalues of A using eig(A)
              -3
              -3
eig(A)  =      2
               2
This shows that p(z)  =  (z-2)^2*(z-3)^2
```

EXERCISE SET 8.3
TECHNOLOGY EXERCISES

T1. Most linear algebra technology utilities do not have a specific command for orthogonally diagonalizing a symmetric matrix, so other commands must usually be pieced together for that purpose. Use the commands for finding eigenvectors and performing the Gram-Schmidt process to find a matrix P that orthogonally diagonalizes the following matrix A. Use your result to factor A as $A = PDP^T$, where D is diagonal.

$$A = \begin{bmatrix} \frac{1}{2} & 0 & \frac{3}{2} & 0 \\ 0 & \frac{1}{2} & 0 & \frac{3}{2} \\ \frac{3}{2} & 0 & \frac{1}{2} & 0 \\ 0 & \frac{3}{2} & 0 & \frac{1}{2} \end{bmatrix}$$

There is no need for the Gram-Schmidt process because $[P,D]$ = eig(A) handles symmetric matrices with no difficulty and high accuracy. When A is symmetric, P is an orthogonal matrix that satisfies $A = PDP^T$. M-file T1_8_3 creates A and provides a symbolic approach too. When A is symbolic, P is no longer an orthogonal matrix and column normalization is necessary to get an orthogonal matrix. The code is written so that you can experiment with small, symmetric matrices of your choice. It is best if the matrices have integer coefficients.

T2. Confirm that the matrix in A in Exercise T1 satisfies its characteristic equation, in accordance with the Cayley-Hamilton Theorem.

A problem with the Cayley-Hamilton Theorem is that you need the coefficients of the characteristic polynomial before calculating the matrix polynomial. They must be known exactly or to high accuracy, and it is generally difficult to obtain this information but not in this case. The code fragments below indicate a symbolic approach, and the M-file T2_8_3 gives a numerical approach.

```
>> As = sym(A) % get symbolic version of A
As =

[ 1/2,   0, 3/2,   0]
[   0, 1/2,   0, 3/2]
[ 3/2,   0, 1/2,   0]
[   0, 3/2,   0, 1/2]

>> syms x % define a symbolic variable
>> p = det(x*eye(4)-As) % characteristic polynomial
```

```
p =

x^4-2*x^3-3*x^2+4*x+4 % this looks encouraging

% Substitute As for x so as to verify the Cayley-Hamilton result
>> disp(subs(p,As))

[ 81/16,       4, 25/16,       4]
[      4, 81/16,       4, 25/16]
[ 25/16,       4, 81/16,       4]
[      4, 25/16,       4, 81/16]

% What happened here is the same thing that happens when
% neophytes first work with matrix polynomials. Since
% x was not part of the constant term, no substitution
% of the identity matrix was made there.
% A corrected version verifies the result.

>> disp(As^4 - 2*As^3 - 3*As^2 + 4*As + 4*eye(4))

[ 0, 0, 0, 0]
[ 0, 0, 0, 0]
[ 0, 0, 0, 0]
[ 0, 0, 0, 0]
```

M-file T2_8_3 returns ancillary information before verifying the Cayley-Hamilton theorem.

```
>> T2_8_3
          1/2     0   3/2     0
            0   1/2     0   3/2
A =       3/2     0   1/2     0
            0   3/2     0   1/2

The coefficients of the characteristic polynomial
are p = [1 -2 -3  4  4].

Using polyrepm, the matrix polynomial p(A)
looks like A^4 - 2A^3 - 3A^2 + 4A + 4I.

polyvalm(p,A) should be zero.
                  0  0  0  0
                  0  0  0  0
polyvalm(p,A) =   0  0  0  0
                  0  0  0  0
```

The polynomial form A^4 - 2A^3 - 3A^2 + 4A + 4I seen in the output was created by a function polyrepm that takes the coefficients in p and returns a formatted expression for the corresponding matrix polynomial including the identity matrix in the constant term. It is included with the other M-files for Chapter 8.

194

T3. Compute e^A for the matrix A in Exercise T1.

Make sure you don't confuse the notation $e^A = \exp(A)$ with the MATLAB operation $\exp(A)$. A matrix exponential is calculated with $\text{expm}(A)$ when A is numeric. They mean different things. M-file T3_8_3 takes a numeric approach and a symbolic approach, and a partial listing of the results follow.

```
>> T3_8_3
            1/2     0     3/2     0
              0    1/2      0    3/2
A =         3/2     0     1/2      0
              0    3/2      0    1/2


This is NOT the matrix exponential of A.
            1.6487          1    4.4817          1  ← note exp(0) here
                 1     1.6487         1     4.4817
exp(A) =    4.4817          1    1.6487          1
                 1     4.4817         1     1.6487


The correct version comes from expm(A).
            3.8785          0    3.5106          0  ← not the same
                 0     3.8785         0     3.5106
expm(A) =   3.5106          0    3.8785          0
                 0     3.5106         0     3.8785
```

T4. Find the spectral decomposition of the matrix A in Exercise T1.

A spectral decomposition can be described by writing $A = PDP^T = \sum_{k=1}^{n} \lambda_k p_k p_k^T$, where p_k is column k of P. This is what is done in T4_8_3, but it is a difficult result to display effectively.

EXERCISE SET 8.4
TECHNOLOGY EXERCISES

T1. Find an orthogonal change of variable that eliminates the cross-product terms from the quadratic form
$$Q = 2x_1^2 - x_2^2 + 3x_3^2 + 4x_4^2 - x_1x_2 - 3x_1x_4 + 2x_2x_3 - x_3x_4$$
and express Q in terms of the new variables.

Generating the matrix A for the quadratic form is tedious. Once the upper triangular part is known, MATLAB primitives can be used to get the symmetric part. If available, the symbolic toolbox can be used to verify the choice of A to get Q. An abbreviated version of T1_8_4 follows.

```
>> T1_8_4
          2    -1/2     0    -3/2
        -1/2    -1      1      0
A =       0      1      3    -1/2
        -3/2     0    -1/2     4
     ***    ***    ***    ***    ***    ***    ***    ***
Here are the results of [P,D] = eig(A);

      -0.15594     0.84538    -0.2481      -0.44661
      -0.96253    -0.10448     0.25022    -0.00069683
P =    0.22062     0.19307     0.92862     -0.22742
      -0.023275    0.48697     0.11621      0.86534


      -1.3102         0          0         0
          0       1.1977         0         0
D =       0          0       3.2069       0
          0          0          0      4.9056


An orthogonal change of variables x = Py changes
the quadratic form Q into

    Q = -1.3102y1^2 + 1.1977y2^2 + 3.2069y3^2 + 4.9056y4^2
```

You can see the eigenvalues of A in Q.

T2. Many linear algebra technology utilities have a command for finding a Cholesky factorization of a positive definite symmetric matrix. The Hilbert matrix

$$A = \begin{bmatrix} 1 & \frac{1}{2} & \frac{1}{3} & \frac{1}{4} \\ \frac{1}{2} & \frac{1}{3} & \frac{1}{4} & \frac{1}{5} \\ \frac{1}{3} & \frac{1}{4} & \frac{1}{5} & \frac{1}{6} \\ \frac{1}{4} & \frac{1}{5} & \frac{1}{6} & \frac{1}{7} \end{bmatrix}$$

is obviously symmetric. Show that it is positive definite by finding its eigenvalues, and then find a Cholesky factorization $A = LL^T$, where L is lower triangular. [**Note**: It is possible that your utility may produce a factorization of the form $A = U^T U$ in which U is upper triangular. If so, let $L = U^T$.] Check your result by computing LL^T.

The command `chol(A)` produces an upper triangular matrix R such that $A = RR^T$ unless A is not positive definite.

```
>> A = [-1 0;0 1]
A =
     -1      0
      0      1
>> R = chol(A)
??? Error using ==> chol
Matrix must be positive definite.
```

Another call avoids the error message and even tells when the algorithm failed.

```
>> help chol

 CHOL   Cholesky factorization.

    [R,p] = CHOL(X), with two output arguments, never produces an
    error message.  If X is positive definite, then p is 0 and R
    is the same as above.   But if X is not positive definite, then
    p is a positive integer.

>> [R,p] = chol(A)
R =
      []
p =
      1
```

The output indicates that the algorithm failed on the first iteration and an empty array is returned because nothing was calculated. This version is an efficient way to test whether a matrix is positive definite – check p.

```
>> T2_8_4
The Hilbert matrix is
          1     1/2    1/3    1/4
        1/2     1/3    1/4    1/5
A =     1/3     1/4    1/5    1/6
        1/4     1/5    1/6    1/7

The Cholesky factor R is
          1       0.5    0.33333        0.25
          0   0.28868    0.28868     0.25981
R = 0           0     0.074536      0.1118
          0         0          0    0.018898

R'*R recovers the Hilbert matrix
                    1     1/2    1/3    1/4
                  1/2     1/3    1/4    1/5
A = R'*R =        1/3     1/4    1/5    1/6
                  1/4     1/5    1/6    1/7
```

EXERCISE SET 8.5
TECHNOLOGY EXERCISES

T1. Find the maximum and minimum values of the quadratic form

$$Q(\mathbf{x}) = x_1^2 + x_2^2 - x_3^2 - x_4^2 + 2x_1x_2 - 10x_1x_4 + 4x_3x_4$$

subject to the constraint $\|\mathbf{x}\| = 1$.

The maximum and minimum can be taken from the eigenvalues of the associated

quadratic form $Q(x) = x^T A x$ because $\lambda_{\min} \le x^T A x \le \lambda_{\max}$ when $\|x\| = 1$.

```
>> T1_8_5
Q(x) = x1^2 + x2^2 - x3^2 - x4^2 + 2x1*x2 - 10x1*x4 + 4x3*x4
The symmetric matrix A for the quadratic form Q = x'*A*x is
        1   1   0  -5
        1   1   0   0
A  =    0   0  -1   2
       -5   0   2  -1

            -5.6817
            -0.80235
eig(A)  =         1
             5.484
```

```
Scan the eigenvalues of A to find the maximum and minimum values of Q
when x is a unit vector.
Qmin = -5.6817
Qmax = 5.484
```

EXERCISE SET 8.6
TECHNOLOGY EXERCISES

T1. (*Singular Value Decomposition*) Most linear algebra technology utilities have a
command for finding the singular value decompositions, but they vary considerably.
Some utilities produce the reduced singular value decomposition, some require that
entries be in decimal form, and some produce U in transposed form, so you will need to
be alert to this. Find the reduced singular value decomposition of the matrix

$$A = \begin{bmatrix} 3 & 0 & \frac{3}{2} \\ 1 & -2 & \frac{5}{2} \\ 1 & -2 & \frac{5}{2} \\ 3 & 0 & \frac{3}{2} \end{bmatrix}$$

and check your answer by multiplying out the factors and comparing the product to *A*.

An "economy size" SVD did not produce a truncated version of *U* because *A* is square. One of the singular values is zero.

```
>> T1_8_6
        3    0   3/2
A =     1   -2   5/2
        1   -2   5/2

After executing [U,S,V] = svd(A)
       -0.49615    0.86824          0
U =    -0.61394   -0.35083   -0.70711
       -0.61394   -0.35083    0.70711


       5.2872         0   0
S =         0    2.4073   0
            0         0   0


       -0.51375    0.79054   -0.33333
V =     0.46447    0.58295    0.66667
       -0.72134   -0.18768    0.66667


                              5.2872
singular values of A =        2.4073
                                   0
1 zero singular value is detected.
Drop the last 1 column(s) of U and V to obtain
the reduced decomposition A = Ur*Sr*Vr', where
       -0.49615    0.86824
Ur =   -0.61394   -0.35083
       -0.61394   -0.35083


       5.2872         0
Sr =        0    2.4073


       -0.51375    0.79054
Vr =    0.46447    0.58295
       -0.72134   -0.18768

As a final check, A = Ur*Sr*Vr' should hold.
               3    0   1.5
Ur*Sr*Vr' =    1   -2   2.5
               1   -2   2.5
Except for roundoff errors, it works.
```

T2. Find the singular values of the matrix

$$A = \begin{bmatrix} 3 & 2 & -5 \\ -6 & -8 & -6 \\ -5 & -5 & 8 \end{bmatrix}$$

by finding the square roots of the eigenvalues of $A^T A$. Compare your results to those produced by the command for computing the singular value decomposition of A.

```
>> T2_8_6
         3   2  -5
A  =    -6  -8  -6
        -5  -5   8
The square roots of the eigenvalues of A'*A are
                0.534778
eig(A'*A).^.5 =  11.0244
                12.8909

The computed singular values of A are
                12.8909
svd(A)  =       11.0244
                0.534778
```

T3. Construct a 3 × 6 matrix of rank 1, and confirm that the rank of the matrix is the same as the number of nonzero singular values. Do the same for 3 × 6 matrices of rank 2 and rank 3.

Choices are displayed when T3_8_6 is executed. MATLAB's `rank` function is used, and only a partial listing is provided.

```
>> T3_8_6
Display matrices A with different ranks.

                        1  1  1  1  1  1
A = ones(3,1)*ones(1,6) = 1  1  1  1  1  1
                        1  1  1  1  1  1
rank(A)  = 1
The singular values of A are
            4.2426
svd(A)  =       0
                0
A is called A1 in the code.
    ***    ***    ***
```

T4. (**MATLAB AND INTERNET ACCESS REQUIRED**). This problem, which is specific to MATLAB, will enable you use singular value decompositions to recreate the baboon pictures in Figure 3 from a scanned image that we have stored on our website for you. The following steps will enable you to produce a rank r approximation of the baboon image:

(1) Download the uncompressed scanned image baboon.bmp from our website http://www.xxxxxx by following the directions posted on the site. As a check, view this image using any program or utility for viewing bitmap images (.bmp files). This image should look like the rank 128 picture in Figure (3).

(2) Use the MATLAB command **graybaboon = imread(`baboon.bmp`)** to assign the pixels in the bitmap image integer values representing their gray levels; the set of gray level integers is named "graybaboon".

(3) Use the MATLAB command **A = double(graybaboon)** to convert the set of gray level integers to a matrix **A** with floating point entries.

(4) Use the MATLAB command **[u,s,v] = svd(A)** to compute the matrices in the singular value decomposition of **A**.

(5) To create a rank r approximation of the baboon image, use appropriate MATLAB commands to form the matrices **ur, sr,** and **vr**, where **ur** consists of the first r columns of **u, sr** is the matrix formed from by first r rows and first r columns of **s**, and **vr** consists of the first r columns of **v**. Use appropriate MATLAB commands to compute the product $(\mathbf{ur})(\mathbf{sr})(\mathbf{vr})^T$ and name it **Ar**.

(6) Use the MATLAB command **graylevelr = uint8(Ar)** to convert the entries of **Ar** to integer gray level values; the matrix of gray level values is named "graylevelr".

(7) Use the MATLAB command **imwrite(graylevelr, `baboonr.bmp`)** to create a bitmap file of the rank r picture named "baboonr.bmp" that can be viewed image using any program or utility for viewing bitmap images.

Use the steps outlined to create and view the rank50, rank20, rank10, and rank 4 approximations in Figure 3.

Nothing is offered for this exercise. Follow the instructions.

EXERCISE SET 8.7
TECHNOLOGY EXERCISES

T1. (*Pseudoinverse*) Some linear algebra technology utilities provide a command for finding the pseudoinverse of a matrix. Determine whether your utility has this capability, and if so, use that command to find the pseudoinverses of the matrix in Example 1.

A pseudoinverse can be computed using the command `pinv`. It is instructive to build a function M-file that also determines a pseudoinverse. That is provided in `pinverse`, and it should be consulted for algorithmic details. T1_8_7 uses both functions and offers confirmatory evidence that they produce the same pseudoinverse.

```
>> T1_8_7
        1   1
A =  0   1
        1   0

A pseudoinverse of A is found using pinv(A).

              0.33333   -0.33333    0.66667
pinv(A)  =  0.33333    0.66667   -0.33333

pinv(A) has the size of the transpose of A.

It should be true that pinv(A)*A = I.

                    1   0
pinv(A)*A  =  0   1

    ***      ***      ***
Use the M-file pinverse and get another version.
                    0.33333   -0.33333    0.66667
pinverse(A)  =  0.33333    0.66667   -0.33333
```

T2. Use a reduced singular value decomposition to find the pseudoinverse of the matrix

$$A = \begin{bmatrix} 1 & 2 & 3 \\ -2 & 1 & -1 \\ 3 & 4 & 7 \end{bmatrix}$$

If your technology utility has a command for finding pseudoinverses, use it to check the result you have obtained.

Computing a pseudoinverse with a reduced singular value decomposition is carried out in T2_8_7. The function M-file `pinverse` records those calculations in case you want to experiment. No output is offered.

T3. The rank of a matrix is the number of linearly independent rows (or columns). In practice, it is difficult to find the rank of a matrix exactly because of roundoff error, particularly for matrices of large size. A common procedure for estimating the rank of a matrix A is to set some small error tolerance ε (depending on the accuracy of the data), and estimate to the rank of A to be the number of singular values of A that are greater than ε. This is called the **effective rank** of A for the given tolerance.

(a) Using a tolerance of $\varepsilon = 10^{-14}$, find the effective rank of the matrix

$$\begin{bmatrix} 16 & 2 & 3 & 13 \\ 5 & 11 & 10 & 8 \\ 9 & 7 & 6 & 12 \\ 4 & 14 & 15 & 1 \end{bmatrix}$$

(b) Use Formula (5) to find the condition number of the matrix

$$A = \begin{bmatrix} -149 & -50 & -154 \\ 537 & 180 & 546 \\ -27 & -9 & -25 \end{bmatrix}$$

(c) Assume the entries in A are exact. How many significant digits are needed for the entries of **b** in order to guarantee three significant digits of accuracy in the solution to the system $A\mathbf{x} = \mathbf{b}$?

All parts are addressed in the M-file T3_8_7. Part (c) is answered by consulting the last two paragraphs of this section in the text. These questions may seem remote, but it is possible to find a 2x2 matrix with integer entries that has rather unusual features. An M-file `funny` provides output for you to consider and only some is shown here.

```
>> funny
% Select two large integers
```

```
u = 9.007199254740991e+15;
v = 9.007199254740992e+15;
u = -1+2^52; % another way to look at it
v = 2^52;
% Create a matrix A whose row reduced form is eye(2)
A = [1 u ;1 v]
A =
     1.000000000000000e+000      4.503599627370495e+015
     1.000000000000000e+000      4.503599627370496e+015
echo off
   u - v = -1
   det(A) = 1
R = rref(A)
R =
     0      1
     0      0
% Based on R = rref(A), det(A) = 0
```

Who do you trust? Read further in the M-file for comments.

T4. Consider the inconsistent linear system $A\mathbf{x} = \mathbf{b}$ in which

$$A = \begin{bmatrix} 1 & 2 & 3 \\ -2 & -3 & -5 \\ 1 & 3 & 4 \end{bmatrix}; \quad \mathbf{b} = \begin{bmatrix} 1 \\ 1 \\ 2 \end{bmatrix}$$

Show that the system has infinitely many least-squares solutions, and use the pseudoinverse of A to find the least squares solution that has minimum norm.

The null space of A is not empty, and that is why the least squares problem has infinitely many solutions.

```
>> T4_8_7
         1   2   3
A =  -2  -3  -5
         1   3   4


         1
b =  1
         2
```

```
The dimension of the null space of A is 3 - rank(A) = 1, and that
is why the least squares problem has infinitely many solutions.
Use the pseudoinverse to obtain a least squares
solution of Ax = b of minimum norm.

                    -2.5758
x = pinv(A)*b =      2.1515
```

204

-0.42424

EXERCISE SET 8.8
TECHNOLOGY EXERCISES

T1. (***Arithmetic operations on complex numbers***) Most linear algebra technology programs have a syntax for entering complex numbers and can perform additions, subtractions, multiplications, divisions, conjugations, modulus and argument determinations, and extractions of real and imaginary parts on them. Enter some complex numbers and perform various computations with them until you feel you have mastered the operations.

The possibility of complex arithmetic is built into much of MATLAB. You can proceed without worrying about complex number or arithmetic and just concentrate on interpreting the results. Mathematics texts present complex numbers as $z = x + iy$, where $i^2 = -1$. The complex conjugate of z is written as $z^* = x - iy$ or $\bar{z} = x - iy$. Some use j instead of i, but MATLAB uses the latter. Both are acceptable. This aspect of complex numbers is addressed in the help files when you type `help i` and read on. The command conj(z) will evaluate the complex conjugate of every element of z and return an array of the same size. The code fragments suggest how complex arithmetic is carried out and displayed.

```
>> z = 2-3i; % 3*i works too
>> disp(z)
   2.0000 - 3.0000i

>> w = 3 - 4*i;
>> disp(w)
   3.0000 - 4.0000i

>> disp(z+w)
   5.0000 - 7.0000i

>> disp(2*z) % rescale by 2
   4.0000 - 6.0000i

>> disp(z/w) % division
   0.7200 - 0.0400i
```

```
>> disp(conj(z)) % complex conjugate of z = 2 -3i
   2.0000 + 3.0000i

>> help conj

 CONJ    Complex conjugate.
    CONJ(X) is the complex conjugate of X.
    For a complex X, CONJ(X) = REAL(X) - i*IMAG(X).
    See also REAL, IMAG, I, J.

>> disp(real(z)) % real part of z
    2

>> disp(imag(z)) % imaginary part of z. Note minus sign
    -3

>> disp(norm(z)) % sometimes called the modulus of z

   3.6056
```

The norm(modulus) of $z = x + iy$ is defined as $|z| = \sqrt{z^{*}z} = \sqrt{x^2 + y^2}$ and it correspond to the length of z when viewed as a vector in the complex plane.

T2. (*Vectors and matrices with complex entries*) For most linear algebra technology utilities, operations on vectors and matrices with complex entries are the same as for vectors and matrices with real entries. Enter some complex numbers and perform various computations with them until you feel you have mastered the operations.

The transpose operation takes on a different form when the matrix is complex. By way of comparison

$$\begin{bmatrix} a_1 \\ a_2 \end{bmatrix}^{T} = \begin{bmatrix} a_1 & a_2 \end{bmatrix} \text{ and in MATLAB } \begin{bmatrix} a_1 \\ a_2 \end{bmatrix}' = \begin{bmatrix} \overline{a_1} & \overline{a_2} \end{bmatrix}$$

The operation A' in MATLAB is technically the conjugate or Hermitian transpose: conjugate each element and take the usual transpose. It makes a difference only when the data is complex. There is standard mathematical notation that reflects this subtle point.

$$\begin{bmatrix} a_1 \\ a_2 \end{bmatrix}^{H} = \begin{bmatrix} \overline{a_1} & \overline{a_2} \end{bmatrix}$$

206

Use transpose(A) to calculate the transpose of *A* without the complex conjugate part. This is useful when working with symbolic matrices because only the transpose is needed for many calculations.

```
>> A = [2-i,1;2-3*i,-2] % A little hard to read
A =

   2.0000 - 1.0000i   1.0000
   2.0000 - 3.0000i  -2.0000

>> disp(size(A))
     2     2

>> disp(A') % complex conjugate transpose
   2.0000 + 1.0000i   2.0000 + 3.0000i
   1.0000            -2.0000

>> disp(transpose(A)) % transpose without conjugate
   2.0000 - 1.0000i   2.0000 - 3.0000i
   1.0               -2.0000

>> disp(inv(A)) % inverse of A
   0.1967 + 0.1639i   0.0984 + 0.0820i
   0.4426 - 0.1311i  -0.2787 - 0.0656i

>> disp(A*inv(A)) % should be the identity
   1.0000 + 0.0000i   0.0000 + 0.0000i
   0.0000 + 0.0000i   1.0000
```

You are expecting to see the identity matrix without any reference to complex numbers, but roundoff errors occurred during the calculations. The current format gives the results to five figures. A more realistic picture is presented when an exponential format is requested and you can see the small entries.

```
>> format short e
>> disp(A*inv(A))

   1.0000e+000 +2.7756e-017i   2.7756e-017 +2.7756e-017i
   5.5511e-017 +1.1102e-016i   1.0000e+000
```

The complex parts are so small they are interpreted as zero for all practical purposes. What is left is the anticipated identity matrix. Such are the vagaries of floating point arithmetic.

The dot product is different when complex vectors are involved. It is symmetric (dot(x,y) = dot(y,x)) when real data is used. It is not when x or y are complex.

```
>> help dot

 DOT   Vector dot product.
    C = DOT(A,B) returns the scalar product of the vectors
    A and B. A and B must be vectors of the same length.
    When A and B are both column vectors, DOT(A,B) is
    the same as A'*B.

>> x = [z;w];
>> y = [w;2*z];
>> disp(x)
   2.0000 -  3.0000i
   3.0000 -  4.0000i

>> disp(y)
   3.0000 -  4.0000i
   4.0000 -  6.0000i

>> disp(dot(x,y))
  54.0000 -  1.0000i

>> disp(dot(y,x)) % complex conjugate of dot(x,y)
  54.0000 +  1.0000i
```

Order is important, and the help file shows that the complex conjugate enters the calculation because of the Hermitian transpose of the first argument: $\mathrm{dot}(A,B) = A^H B$.

T3. Perform the computations in Examples 1 and 2.

A commented M-file T3_8_8 gives the details and a partial listing follows.
```
>> T3_8_8
Perform computations in Examples 1 and 2.

        3+1i
v =     0-2i
        5+0i

        1+1i   0-1i
A =     4+0i   6-2i

            3-1i
conj(v) =   0+2i
            5+0i
```

An almost identical symbolic version is found in T3sym_8_8.

T4. (a) Show that the vectors

$$\mathbf{u}_1 = (i, i, i), \ \mathbf{u}_2 = (0, i, i), \ \mathbf{u}_3 = (i, 2i, i)$$

are linearly independent.

(b) Use the Gram-Schmidt process to transform $\{\mathbf{u}_1, \mathbf{u}_2, \mathbf{u}_3\}$ into an orthonormal set.

Output of T4_8_8 chronicles the details for this exercise. The QR-decomposition using qr is Gram-Schmidt in disguise. Just remember that the Hermitian transpose is understood when input data is complex.

```
>> T4_8_8
The three vectors in matrix form are
      0+1i   0+0i   0+1i
U =   0+1i   0+1i   0+2i
      0+1i   0+1i   0+1i
The three column vectors are linearly
independent because rank(U) = 3.
The results of Gram-Schmidt are found in the
columns of Q when [Q,R] = qr(U).

      0-0.57735i   0-0.8165i     0+0i
Q =   0-0.57735i   0+0.40825i    0+0.70711i
      0-0.57735i   0+0.40825i    0-0.70711i

As a check, Q'*Q should be the identity.
          1   0   0
Q'*Q =    0   1   0
          0   0   1
The columns of Q constitute an orthonormal set.
```

The computed value of $Q^H Q$ is not quite the identity matrix because of roundoff errors. Execute the next lines to see a more accurate indication.

```
>> format short e
>> disp(QtQ)
```

T5. Determine whether there exist scalars c_1, c_2, and c_3 such that

$$c_1(i, 2 - i, 2 + i) + c_2(1 + i, -2i, 2) + c_3(3, i, 6 + i) = (i, i, i)$$

Except for the complex numbers, this is the familiar problem of solving a linear system $Ac = b$, where the columns of A come from the vectors in the left side of the equation.

The MATLAB command $c = A \backslash b$ works well with complex vectors too, and the data is in the M-file T5_8_8.

T6. Find the eigenvalues and bases for the eigenspaces of

$$A = \begin{bmatrix} 1 & 2 & 1 \\ -2 & 1 & 0 \\ 1 & 0 & 1 \end{bmatrix}$$

The function eig is designed to produce complex eigenvalues, and the corresponding eigenspaces are the spans of the individual columns of P, where [P,D] = eig(A). The M-file T6_8_8 offers hints. Eigenvalues only are returned with a command evals = eig(A).

T7. Factor $A = \begin{bmatrix} -1-\sqrt{3} & 2\sqrt{3} \\ -\sqrt{3} & -1+\sqrt{3} \end{bmatrix}$ as $A = PCP^{-1}$, where C is of Form (19).

There are two approaches, symbolic and numeric. The M-file T7_8_8 takes a symbolic approach and actually starts with the more general matrix

$$A = \begin{bmatrix} -1-u & 2u \\ -u & -1+u \end{bmatrix}$$

and later substitutes $u = \sqrt{3}$ to obtain the result. The M-file T7_8_8 contains several techniques for manipulating symbolic variables when complex numbers are present.

EXERCISE SET 8.9
TECHNOLOGY EXERCISES

T1. Find a matrix P that unitarily diagonalizes the matrix A in Example 2.

The matrix is

$$A = \begin{bmatrix} 1 & i & 1+i \\ -i & -1 & 1 \\ 1-i & 1 & 0 \end{bmatrix}$$

and the command [P,D] = eig(A) gives an invertible matrix P that diagonalizes A. The columns of P are always normalized to be unit vectors, and consequently it is a unitary matrix because A is Hermitian. Output from T1_8_9 describes and displays the calculations that verify these claims.

T2. Find the eigenvalues and bases for the eigenspaces of the Hermitian matrix and confirm that they have the properties stated in Theorems 8.9.4 and 8.9.5.

$$A = \begin{bmatrix} 3 & 3+3i \\ 3-3i & 5 \end{bmatrix}$$

T2_8_9 generates this data for the matrix A, and the eigenvalues are real because A is Hermitian.

```
>> T2_8_9
        3+0i    3+3i
A =     3-3i    5+0i

A is Hermitian because

            0   0
A - A' =    0   0

Use [P,D] = eig(A) to get eigenvalues and vectors.

                    -0.3589
eigenvalues of A =   8.3589

The corresponding eigenvectors are the columns of

        0.5544+0.5544i   0.43891+0.43891i
P =    -0.62072+0i       0.78403+0i
```

T3. Approximate the eigenvalues and bases for the eigenspaces of the Hermitian matrix to two decimal places.

$$A = \begin{bmatrix} 3 & -3+2i & 2-2i \\ -3+2i & 1 & 4+2i \\ 2+2i & 4-2i & -4 \end{bmatrix}$$

Standard output using `eig` provides this information, and the short display format gives the eigenvalues and eigenvectors to at least two decimal places. Output of T3_8_9 is not included here.

T4. (CAS) Find the eigenvalues and bases for the eigenspaces of the Hermitian matrix

$$A = \begin{bmatrix} 1 & a+bi \\ a-bi & 1 \end{bmatrix}$$

The same command eig works with small matrices and the symbolic results are easy to check.

```
>> T4_8_9
      [     1, a+i*b]
A =   [a-i*b,     1]

                        [1+(b^2+a^2)^(1/2)]
eigenvalues of A =      [1-(b^2+a^2)^(1/2)]

The corresponding eigenvactors are the columns of
        [1/(b^2+a^2)^(1/2)*(a+i*b),  (-a-i*b)/(b^2+a^2)^(1/2)]
P =   [                         1,                         1]
```

EXERCISE SET 8.10
TECHNOLOGY EXERCISES

The main tools for solving systems of differential equations are symbolic in nature

T1. The problem is to discuss the solutions a system of differential equations

$$\begin{aligned} y_1' &= 3y_1 + 2y_2 + 2y_3 \\ y_2' &= y_1 + 4y_2 + y_3 \\ y_3' &= -2y_1 - 4y_2 - y_3 \end{aligned}$$

(a) Find a general solution of the system.

(b) Find the solution that satisfies the initial conditions

$$y_1(0) = 0, y_2(0) = 1, y_3(0) = -3$$

Presented here are four strategies in MATLAB when the Symbolic Toolbox is available. One is numeric and three are symbolic. Two use eigenvalues and one uses the matrix exponential. The only unusual approach involves a symbolic function called dsolve. Type doc dsolve for online help details or help dsolve for command window help.

```
>> de = 'Dx = -a*x'; % D indicates a derivative
>> sol = dsolve(de) % return a symbolic solution expression
sol =

C1*exp(-a*t) % a Is unspecified at this point

>> sol2 = dsolve(de,'x(0) = 1') % include initial conditions
sol2 =

exp(-a*t)
```

The output of T1_8_10 is too extensive to give in full, but a partial listing is offered.

```
>> T1_8_10
The differential equation is dy/dt = Ay and
         3   2   2
A =      1   4   1
        -2  -4  -1
and the initial conditions are
          0
y(0) =    1
         -3
    ***   ***   ***   ***   ***


The solution expression is
    y(t) = p1*exp(2t)*(2) + p2*exp(1t)*(-4) + p3*exp(3t)*(1)

where p1, p2, and p3 are the columns of the matrix
        -2  -1   0
P =      1   0  -1
         0   1   1


    ***   ***   ***   ***   ***


The symbolic solution using the matrix exponential is
         [-4*exp(2*t)+4*exp(t)]
y(t) =   [ 2*exp(2*t)-exp(3*t)]
         [   exp(3*t)-4*exp(t)]
```

T2. The electrical circuit in an accompanying figure in the text, called a *parallel LRC circuit*, contains a resistor with resistance R ohms (Ω), an inductor with inductance L henries (H), and a capacitor with capacitance C farads (F). It is shown in electrical circuit theory that the current I in amperes (A) though the inductor and the voltage drop V in volts (V) across the capacitor satisfy the system of differential equations

$$\frac{dI}{dt} = \frac{V}{L}$$

$$\frac{dV}{dt} = -\frac{I}{C} - \frac{V}{RC}$$

where the derivatives are with respect to the time t. Find I and V as functions of t if $L = 0.5$ H, $C = 0.2$ F, $R = 2$ Ω, and the initial values of V and I are $V(0) = 1$ V and $I(0) = 2$ A.

A symbolic approach is used for this problem because the occurrence of complex eigenvalues complicates a description of a solution expression. Similar comments can be made about a numeric approach. This is not the way scientists usually analyze circuit problems. The matrix presentation is

$$\begin{bmatrix} \dfrac{dI}{dt} \\ \dfrac{dV}{dt} \end{bmatrix} = \begin{bmatrix} 0 & 1/L \\ -1/C & -V/RC \end{bmatrix} \begin{bmatrix} I \\ V \end{bmatrix}$$

and a matrix exponential calculation gives the explicit solution without much user effort. The ouput of T2_8_10 is presented, and the file has several comments to help lead you through the details.

```
>> T2_8_10
The differential equation is dy/dt = Ay, where
     [I]
y = [V]


     [ 0,    2]
A = [-5,  -5/2]


The eigenvalues of A are complex, and this causes
```

trignometric functions to be part of a real solution.

Review Euler's formula in Appendix B.

For initial conditions

$$y(0) = \begin{matrix} 2 \\ 1 \end{matrix}$$

the solution is the long expression

```
I(t) = [2*exp(-5/4*t)*cos(3/4*t*15^(1/2))+2/5*15^(1/2)*exp(-5/4*t)*sin(3/4*t*15^(1/2))]

V(t) = [-15^(1/2)*exp(-5/4*t)*sin(3/4*t*15^(1/2))+exp(-5/4*t)*cos(3/4*t*15^(1/2))]
```

The font size is reduced for display purposes.

Command Summary

svd   --- svd(A) returns singular values and singular vectors of A

cond   --- cond(A) estimates the condition number of A

polyrepm   --- produces a character representation of a matrix polynomial

pinv   --- pseudoinverse of a matrix

dsolve   --- symbolic function for solving a differential equation

chol   --- cholesky decomposition of a positive definite matrix

Chapter 9

General Vector Spaces

EXERCISE SET 9.2
TECHNOLOGY EXERCISES

Exercises in this section use some calculus. Integrals must be evaluated and there are two ways to proceed: symbolic or numeric. Numerical integration is carried out using a function called quadl, "quad-L", and the function to be integrated and the finite limits must be supplied. A simple example illustrates quadl and a subtlety.

```
>> f = 'x^3'; % function to be integrated
>> r = quadl(f,-1,1); % integrate over [-1,1] to get 0
??? Error using ==> inlineeval
Error in inline expression ==> x^3
??? Error using ==> ^
Matrix must be square.
```

The difficulty is in the specification of the function f. During execution, the function quadl evaluates f at several points, and x^3 suggests the third power of a matrix in MATLAB notation. MATLAB treats x as a matrix at some point and a slightly different approach is required.

```
>> f = 'x.^3'; % apply the cube operator to each element
>> r = quadl(f,-1,1);
>> disp(r)
  1.3878e-017
```

The expected answer is zero, and roundoff errors together with the strategies used in quadl keep the result from being exact. This is not unusual and it is generally nothing to be concerned about in most problems. A symbolic approach uses the function int, and it is mandatory to consult help int for details on its use.

```
>> t = int(f,-1,1);
??? Error using ==> sym/sym (char2sym)
x.^3 is not a valid symbolic expression.
```

```
>> f = 'x^3'; % use the more natural form
>> t = int(f,-1,1);
>> disp(t) % anticipated result
        0
```

This shows that `quadl` and `int` are really different.

The most interesting part of T1 is not the calculation of the Fourier coefficients, but the graphs of the resulting Fourier approximations. Both symbolic and numeric approaches are offered. When T1_9_2 is executed and you have the symbolic toolbox installed, a small menu appears in the upper left corner of your screen. It gives you a choice of a numeric or symbolic strategy. Select one and proceed. There is no choice if you don't have the symbolic toolbox. After making your selection, another menu appears and it gives you an opportunity to select the different order Fourier approximations. It is probably the most extensive M-file in this technology manual and it only uses elementary MATLAB primitives.

T1. Find the Fourier approximations of orders 1, 2, 3, and 4 to the function $f(x) = x^2$, and compare the graphs of those approximations to the graph of f over the interval $[0, 2\pi]$.

The Fourier approximation of order k is $f_k(x) = a_0/2 + \sum_{j=1}^{k} a_j \cos(jx) + b_j \sin(jx)$ and the coefficients are specified as

$$a_j = \frac{1}{\pi} \int_0^{2\pi} f(x)\cos(jx)dx, \quad b_j = \frac{1}{\pi} \int_0^{2\pi} f(x)\sin(jx)dx$$

The solution strategy for this problem is dictated by the method used to carry out the integrations. Once the coefficients are known, the most interesting thing to do is to graph the result and compare it with the function f. Fourier coefficients are used for more sophisticated purposes in an advanced course, and those issues are not addressed in this manual. A snippet of MATLAB code gives you a glimpse of the main idea and a figure to examine. A numeric approach is used here and corresponding symbolic code is found in T1sym_9_2. The last character in quadl is not the numeral 1.

```
>> f = 'x.^2';
>> a0 = (1/pi)*quadl(f,0,2*pi); % quadl does numerical integration
>> f1 = strcat(f,'.*cos(x)');% f(x)*cos(x) in string form
>> a1 = (1/pi)*quadl(f1,0,2*pi)/pi;
>> g1 = strcat(f,'.*sin(x)');
>> b1 = (1/pi)*quadl(g1,0,2*pi);
>> x = linspace(0,2*pi); % data for plotting
>> fx = x.^2; % reference curve
>> fapprox = a0/2 + a1*cos(x) + b1*sin(x); % first order approximation
>> plot(x,fx,x,fapprox) % plot both
>> axis([0,2*pi,-5,40]) % for figure placement
>> legend(f,'first order Fourier',2) % legend in upper left corner
```

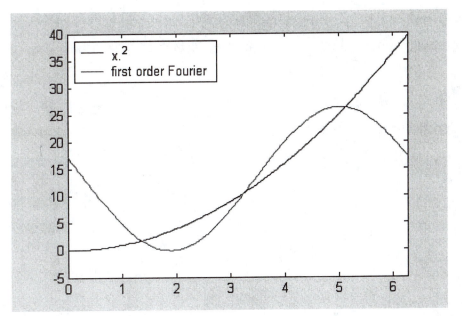

The figure above gives some idea of the user interface without the menu. Other parts of the problem consist of more approximations. One problem with just writing more code is that it is difficult to switch from one approximation to the other. To this end, a more ambitious plan is developed to provide an elementary interactive program that enables you to compare approximations. Two approaches are possible because the Fourier coefficients are given by integrals. One approach is symbolic and the other is strictly numeric. Both are offered. The M-file T1_9_2 coordinates this approach and it, in turn,

219

calls two other M-files T1sym_9_2 or T1num_9_2 that carry out the symbolic or numeric computations, respectively. The latter two M-files stand can be executed separately and you can skip T1_9_2 entirely. An attractive feature of these supporting M-files is that you can edit them and prescribe your own function to be approximated. Read the first few lines of T1sym_9_2 or T1num_9_2 for instructions and suggestions.

T2. In Example 3 of Section 7.8 we found the least squares solution of the linear system

$$
\begin{array}{rcrcl}
x_1 & - & x_2 & = & 4 \\
3x_1 & + & 2x_2 & = & 1 \\
-2x_1 & + & 4x_2 & = & 3
\end{array}
$$

Use the result in Exercise P6 to find the least squares solution of this system with respect to the weighted Euclidean inner product $\langle \mathbf{u}, \mathbf{v} \rangle = 3u_1 v_1 + 2u_2 v_2 + u_3 v_3$

This solution follows the plan laid out in Exercise P6. When written out, the inner product defines the weight matrix M as

$$
M = \begin{bmatrix} 3 & 0 & 0 \\ 0 & 2 & 0 \\ 0 & 0 & 1 \end{bmatrix}
$$

because

$$
\langle \mathbf{u}, \mathbf{v} \rangle = 3u_1 v_1 + 2u_2 v_2 + u_3 v_3 = \begin{bmatrix} u_1, u_2, u_3 \end{bmatrix} \begin{bmatrix} 3 & 0 & 0 \\ 0 & 2 & 0 \\ 0 & 0 & 1 \end{bmatrix} \begin{bmatrix} v_1 \\ v_2 \\ v_3 \end{bmatrix}
$$

A weighted least squares solution is found by solving the modified normal equations $A^T M A x = A^T M b$, where

$$
A = \begin{bmatrix} 1 & -1 \\ 3 & 2 \\ -2 & 4 \end{bmatrix}, b = \begin{bmatrix} 4 \\ 1 \\ 3 \end{bmatrix}
$$

T2_9_2 carries out this calculation and compares it with the standard least squares solution found by solving the normal equations $A^T A y = A^T b$. These solutions are different, and only the context of the problem can provide a meaningful interpretation of the weighted case. This point is taken up in more advanced courses.

```
>> T2_9_2
Solutions
    weighted   unweighted
    0.4748     0.1789
    0.1306     0.5018
```

T3. (CAS) Let W be the subspace of $C[-1, 1]$ spanned by the linearly independent polynomials 1, x, x^2, and x^3. Show that if the Gram-Schmidt process is applied to these polynomials using the integral inner product

$$\langle f, g \rangle = \int_{-1}^{1} f(x)g(x)\, dx$$

then the resulting orthonormal basis vectors for W are

$$\frac{1}{\sqrt{2}}, \quad \sqrt{\frac{3}{2}}\, x, \quad \frac{1}{2}\sqrt{\frac{5}{2}}\left(3x^2 - 1\right), \quad \frac{1}{2}\sqrt{\frac{7}{2}}\left(5x^3 - 3x\right)$$

These are called **Legendre polynomials** in honor of the French mathematician Adrien-Marie Legendre (1752 – 1833 who first recognized their importance in the study of gravitational attraction. [**Note**: To solve this problem you will probably have to piece together integration commands and commands for applying the Gram-Schmidt process to functions.]

A faithful translation of the Gram-Schmidt process is carried out in the M-file T3_9_2 using the symbolic toolbox. It follows the description given in the text, and each step is explained in detail. Symbolic integrations for the inner product are carried out using int. The M-file is commented so that you can follow the calculations. At the end, additional computations are carried out to verify that the polynomials are indeed orthogonal and of unit length using the integral formula for the inner product. Graphs of the Legendre polynomials are included too, and the legend on the figure makes more sense when viewed in color. Additional comments are added to the command window output below because the reporting style of radicals is not transparent.

```
>> T3_9_2 % Use Gram-Schmidt to obtain Legendre polynomials
   Legendre Polynomials
```

p0 = 1/2*2^(1/2) $\leftrightarrow \dfrac{1}{\sqrt{2}}$

p1 = 1/2*x*6^(1/2) $\leftrightarrow \sqrt{\dfrac{3}{2}}x$

p2 = 1/4*(3*x^2-1)*10^(1/2) $\leftrightarrow \dfrac{1}{2}\sqrt{\dfrac{5}{2}}\left(3x^2-1\right)$

p3 = 1/4*x*(5*x^2-3)*14^(1/2) $\leftrightarrow \dfrac{1}{2}\sqrt{\dfrac{7}{2}}\left(5x^3-3x\right)$

```
*** Orthogonality Check
<p0,p1> = 0
<p0,p2> = 0
<p0,p3> = 0
<p1,p2> = 0
<p1,p3> = 0
<p2,p3> = 0

***    Unit Vector Check    ***
<p0,p0> = 1
<p1,p1> = 1
<p2,p2> = 1
<p3,p3> = 1
```

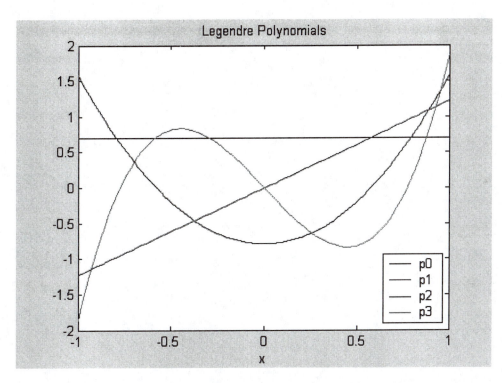

Legendre graphs look more interesting when viewed in color.

Command Summary

quadl   ---- numerical integration function
int   ---- symbolic integration function
char   --- for us, char(xsym) converts a symbolic expression to character form.

Help files should be consulted for a discussion of these commands and other commands used but not mentioned here. The command `char` is an extensive function that can only be appreciated by typing `doc char` or `help char`.

<u>NOTES</u>

NOTES

NOTES

NOTES

NOTES

NOTES

NOTES

NOTES

NOTES